Leading the Way to Competitive Excellence

Also available from ASQ Quality Press

Let's Work Smarter, Not Harder: How to Engage Your Entire Organization in the Execution of Change
Michael Caravatta

Experiences of a CEO: How to Make Continuous Improvement in Manufacturing Succeed for Your Company
Hank McHale

The Change Agents' Handbook: A Survival Guide for Quality Improvement Champions
David W. Hutton

The Transition to Agile Manufacturing: Staying Flexible for Competitive Advantage
Joseph C. Montgomery and Lawrence O. Levine, editors

Quality, Safety, and Environment: Synergy in the 21st Century
Pascal Dennis

To request a complimentary catalog of publications, call 800-248-1946.

Leading the Way to Competitive Excellence

The Harris Mountaintop Case Study

William A. Levinson, editor

ASQ Quality Press
Milwaukee, Wisconsin

Leading the Way to Competitive Excellence: The Harris Mountaintop Case Study
William A. Levinson, editor

Library of Congress Cataloging-in-Publication Data
    Leading the way to competitive excellence: the Harris Mountaintop
case study / William A. Levinson, editor.
        p.   cm.
    Includes bibliographical references and index.
    ISBN 0-87389-376-X (alk. paper)
    1. Harris Semiconductor (Firm)   2. Semiconductor industry—United
States—Management—Case studies.   I. Levinson, William A., 1957–   .
HD9696.S44H375   1998
338.7′62138152′0973—dc21   97-17504
                                                               CIP

Trademark Acknowledgment
Many of the designations used by manufacturers and sellers to distinguish their products are claimed as trademarks. Where these designations appear in this book and ASQ Quality Press was aware of a trademark claim, the designations have been printed in initial caps.

10 9 8 7 6 5 4 3 2

ISBN 0-87389-376-X

Acquisitions Editor: Roger Holloway
Project Editor: Jeanne W. Bohn

ASQ Mission: To facilitate continuous improvement and increase customer satisfaction by identifying, communicating, and promoting the use of quality principles, concepts, and technologies; and thereby be recognized throughout the world as the leading authority on, and champion for, quality.

Attention: Schools and Corporations
ASQ Quality Press books, videotapes, audiotapes, and software are available at quantity discounts with bulk purchases for business, educational, or instructional use. For information, please contact ASQ Quality Press at 800-248-1946, or write to ASQ Quality Press, P.O. Box 3005, Milwaukee, WI 53201-3005.

For a free copy of the ASQ Quality Press Publications Catalog, including ASQ membership information, call 800-248-1946.

Printed in the United States of America

 Printed on acid-free paper

American Society for Quality

Quality Press
611 East Wisconsin Avenue
Milwaukee, Wisconsin 53202

Contents

CHAPTER 1

Introduction

William A. Levinson

> *Think through opportunities and understand what it takes*
> *to win—not just play.*
>
> —John Garrett, President, Harris Corporation,
> Semiconductor Sector

In Eliyahu Goldratt and Jeff Cox's *The Goal* (1992), a fictional company uses the Theory of Constraints to achieve a phenomenal turnaround. In the movie *Gung Ho,* a Japanese company buys an American automobile factory. The Western and Asian cultures clash, but the company finally achieves a similar turnaround.

This book is about a real factory that, by choosing to remake itself, changed its fortunes remarkably. This book offers our insights and experiences to those who wish to achieve similar results. We offer a set of principles and ideas, and Mountaintop's experience shows them at work. We encourage the reader to adapt the principles to his or her particular situation. Managerial or professional employees in any manufacturing or service activity should benefit from Mountaintop's experiences.

This book discusses both the hard (technical) and soft (human behavioral) aspects of quality. These include the following:

- Cultural change and social factors
- Teaming and organizational development

- Self-directed work teams (SDWTs) or autonomous work groups
- Total productive maintenance (TPM), and 5S-CANDO
- Integrated yield management (IYM)
- Synchronous flow manufacturing (SFM), which is similar to just-in-time (JIT)
- ISO 9000 and QS-9000
- Statistical process control (SPC) and industrial statistics
- The Internet

These programs and techniques are not, however, separate and independent tools. A common mistake is to pick the quality technique of the year (or month) and expect miraculous results. This is like expecting to complete a major construction project or repair job with only a hammer. Excellence comes from mutually supporting programs and activities, not isolated ones. The book's chapters frequently refer to other chapters and quality programs. The mutual support and synergy of Mountaintop's quality and productivity improvement activities is a recurring theme.

Commitment to quality and excellence must pervade the organization at every level. Average organizations will do what they must to get an ISO 9000 certificate. Excellent organizations view ISO 9000 as a framework and guide for self-assessment and continual improvement. They see the standard as a welcome quality improvement tool, not a costly and time-consuming requirement. A standard can drive creative and innovative thinking about the company's quality system.

Principles, Not Recipes

"If the only tool you have is a hammer, everything looks like a nail." This book's goal is to teach principles, not recipes. Principles and ideas are valuable guides for independent thought, but they are not substitutes for it. The best technical and human management tools, and their application, will depend on your industry and its environment: cultural, economic, and physical.

No single tool or approach works in all situations. We should observe others, learn how they achieve results, and see what we can apply to our situation. We must select only the applicable tools, and we must adapt them according to our needs. Fit the quality management or human resource tool to the situation; don't try to make the situation fit the tool.

Figure 1.1 shows the basic idea. The triangle is the diagram for a quality management system, and we can adapt it to this discussion. Immutable, unyielding principles are at the top. Readers of Stephen R. Covey's *Principle-Centered Leadership* (1991) will recognize them as the organizational compass. Covey talks about the "soft side" of management: principles for leading people and organizations. There also are principles for the "hard side," including production control. A compass always works, no matter who or where its user is. Covey also discusses maps and their limitations. Maps work only in specific places and situations, but a compass works anywhere.

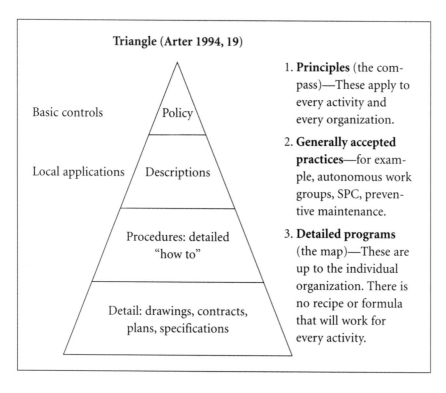

Triangle (Arter 1994, 19)

Basic controls

Policy

Local applications

Descriptions

Procedures: detailed "how to"

Detail: drawings, contracts, plans, specifications

1. **Principles** (the compass)—These apply to every activity and every organization.

2. **Generally accepted practices**—for example, autonomous work groups, SPC, preventive maintenance.

3. **Detailed programs** (the map)—These are up to the individual organization. There is no recipe or formula that will work for every activity.

Figure 1.1. Hierarchy of principles, general practices, and specific practices.

Goldratt's Theory of Constraints was indispensable in achieving huge productivity increases at Mountaintop. Here is the basic principle: Throughput, or delivery of goods or services, is the only thing that earns money. This applies to every organization that offers goods or services. Generally accepted practices include JIT manufacturing, kanban, and SFM. They are all techniques for improving throughput and reducing inventory, and there are many similarities between them. We use SFM, and we have had good results with it. We do not pretend to have a recipe for maximizing throughput and minimizing operating expenses in every industry. Our production control system might not be the best one for your activity. You must, however, follow the underlying principle: A business' goal is to make money, now and in the future. Beware of performance measurements that do not promote throughput, or that even undermine it.

Here is another principle: Ensure quality by controlling the process. SPC applies to manufacturing processes that make discrete items. Chemical factories use automatic controllers that are suitable for continuous processes. The chemical factory and the discrete process follow the same principle: Control the process. They diverge, however, at the generally accepted practice for doing it. Activities that use SPC then face further choices. Should they use traditional Shewhart charts, exponentially weighted moving averages, or cumulative sum? They should use whatever is most effective for their specific processes. This book offers the compass and shows some generally accepted practices.

"Do as I say, not as I do" has obvious deficiencies. Its converse—"Do exactly as I do, imitate me in every detail"—also has problems. It is Taylorism at the managerial or executive level: "Leave your brain at the factory gate." Blind adherence to dogmatic recipes is not a formula for success. There is no formula for success! Instead, look at what we did, pay close attention to the underlying ideas and principles, adapt them to your situation, and improve on them!

Quality Comes from Systems, Not Isolated Programs

A quality system is not a checklist of programs or activities. It is a set of mutually supporting, synergistic programs and activities.

Cohen and Gooch (1991, 51) explain this principle in a military context. Why was Japan's air raid on Pearl Harbor (7 December, 1941) so successful? The United States had all the elements of a powerful air defense system: anti-aircraft guns, fighter planes, and even radar. Radar, a British invention, was not even available to Japan in 1941. The air defense elements, however, did not work together or support each other. In contrast, close cooperation between English radar stations and the Royal Air Force helped win the Battle of Britain.

In 334 B.C.E., Alexander the Great led an invasion force across the Hellespont into Asia Minor. The Persians had three times as many ships as the Greeks, and their sailors were better. A determined attack might have stopped Alexander's invasion before it started. However, the Macedonians did not see even a single enemy warship during the crossing. "Coordinated strategy could not be called the Persian High Command's strongest point" (Green 1991, 167). Again, the system element was there, but the system wasn't.

Quality comes from people who use the programs and techniques as part of a quality system. Stephen R. Covey (1991, 265) says,

> *Deming's "14 Points" are more than a mere checklist of things to do to achieve Total Quality. These points are integrated, interdependent, and holistic. They must be viewed and applied as an interrelated system of paradigms, processes, and procedures—a complete framework of management and leadership harnessed to achieve maximum effectiveness and quality of product and service from the people constituting the enterprise.*

During the 1980s, many companies viewed SPC as a miraculous Japanese success secret. (SPC is actually an American invention that the Japanese adopted enthusiastically.) They thought they could get instant quality by putting SPC charts on every process. They did not understand how the charts worked, and the operators did not understand their purpose. The charts looked impressive on the factory walls, but they did not improve quality. Hradesky (1987, 119) refers to such charts as *wallpaper*. The technique was present, but it did not fit into an overall system.

Table 1.1 shows the interaction of Mountaintop's quality programs.

Table 1.1. Mutually supporting quality system elements.

	Quality System Element
Interaction with:	Self-directed teams, teaming
Cultural change, social factors	The organization's members must accept the SDWT concept and make it work.
ISO 9000, QS-9000	Teams have to make them work. Frontline workers are the key players, and they must know why the programs are important.
Synchronous flow manufacturing	Teams improve yields, reduce scrap. Yield improvement and scrap reduction are especially important after the constraint.
Total productive maintenance, 5S-CANDO	Teams handle preventive maintenance, 5S-CANDO. Team members must acquire the necessary skills to maintain the equipment.
IYM, SPC, design of experiments (DOE), calibration	Teams can take responsibility for SPC.
The Internet	Mountaintop's computer network helps people share information. Some companies have intranets, or internal networks. Harris will use the intranet to let teams post their meeting times, minutes, and projects.
Summary	"Frontline people make it work."
	Cultural change, social factors
ISO 9000, QS-9000	ISO 9000 and QS-9000 preclude the "old way" of doing business. Everyone must buy in to make them work.
SFM	Discard old paradigms about "efficiencies," etc. Don't make inventory to "keep workers busy" or "make the numbers" (production quotas). Recognize the difference between the cost accounting world and the production world.
TPM, 5S-CANDO	The techniques require acceptance and buy-in.
IYM, SPC, DOE	The techniques require acceptance and buy-in.
The Internet	Recognize its value and use it. It will change the way everyone does business.
Summary	"Organizations must accept change to survive and prosper."
	ISO 9000, QS-9000
SFM	Supports QS-9000 goals for continuous (productivity) improvement.
TPM, 5S-CANDO	QS-9000 requires preventive maintenance.

Table 1.1. *Continued.*

ISO 9000, QS-9000 (cont'd.)	
IYM, SPC, DOE	The standards require statistical methods, and the methods support continuous improvement activities.
The Internet	Information resources are available to help meet requirements.
Summary	"ISO 9000 and QS 9000 are valuable guides for continuous improvement of the quality system."
Synchronous flow manufacturing	
TPM, 5S-CANDO	SFM makes the need for TPM more urgent. Prevent down-time, especially at the constraint. Focus TPM efforts on the constraint.
IYM, SPC, DOE	Improve/protect yields, especially at and after the constraint.
The Internet	Online information sources are available.
Summary	"The goal is to make money." We achieve this by delivering products or services. Misleading measurements cause suboptimization.
Total productive maintenance, 5S-CANDO	
IYM, SPC, DOE	TPM suppresses problems that can reduce throughput yield and supports IYM.
The Internet	
Summary	"Suppress friction and win." *Friction* is Carl von Clausewitz's (1976) term for seemingly minor events whose cumulative effects result in failure.
SPC, DOE, calibration	
The Internet	There is a lot of information to help with problem solving and improvements. Links to NIST, other calibration support.
Summary	Calibration: "If you can't measure it, you can't control it."

Commitment to Quality Must Pervade the Organization

Many companies consider ISO 9000 certification an expensive, painful requirement. They want to get the certificate so they can do business in Europe. They want to pass QS-9000 so they can sell to Ford, Chrysler, and General Motors. These companies are like people who think that getting a black belt will make them safe from criminals. The belt does not protect its wearer from anything, but the process of earning it does. Suppose that a company throws a quality system together and manages to earn an ISO 9000 certificate. If the company then stops working on the system (until the next audit), its quality will not improve.

Now consider a company that sees ISO 9000 as a process of continuous improvement. It will improve its productivity and quality, reduce its costs, and satisfy its customers. The process of diligently earning and maintaining certification changes ISO 9000 from an expensive annoyance into a money-making tool (Scotto 1996). Harris Semiconductor's training stresses that ISO 9000's purpose is to help ensure quality and improve productivity. Every employee must understand the program's purpose and his or her role in making it work.

CHAPTER 2

An Introduction to Harris Semiconductor

Raymond Ford

Background

The story began on December 1, 1988, when Harris Semiconductor bought the Mountaintop plant from General Electric/RCA/Intersil. The plant makes discrete power semiconductors for the automotive industry and power control applications. The competitive environment is a commodity market, where price is the principal consideration.

In 1991, the plant was losing money and was facing shutdown. Management had reduced the workforce from 3100 to 500 during the past 10 years, and labor relations were strained. The plant had to start earning a profit, or Harris would close it.

This situation, however, set the stage for a radical turnaround in the plant's fortunes. Had the plant been marginally successful, there would have been no enthusiasm for change. There is a legend that one can cook a frog by putting it in water and heating the water slowly. By the time the frog recognizes the need to change its situation, it is dinner. If one throws the frog in boiling water, it will jump out because it knows immediately that it has a problem. The management team and the union recognized the need for profound changes in the plant's business practices and culture.

Mountaintop made the necessary changes, and Figure 2.1 shows the bottom line. (For confidentiality reasons, the income is in arbitrary units. The fiscal year runs from July through June.) This book will show what we did to achieve these results and how we did it.

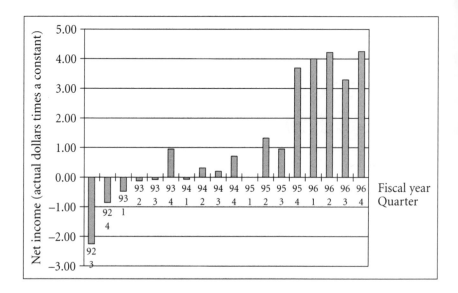

Figure 2.1. Harris Mountaintop net income (in arbitrary units).

Figure 2.2 shows Harris Semiconductor's customer satisfaction index. The measurement is the gap between customer expectations and Harris' performance; lower is better.

Importance of Cultural Change

We cannot overemphasize the importance of cultural change. It is not enough to merely change the business practices. Organizational culture is "a set of shared values and beliefs that guide the behavior of organization members" (Schermerhorn, Hunt, and Osborn 1985, 372).

Stephen Covey (1991) compares formal procedures to maps, and principles to a compass. A map is useful only in familiar territory, while a compass guides its user in new and unfamiliar situations. Management by procedures is characteristic of mechanistic, tightly controlled organizations. Organic, adaptive, and dynamic organizations must rely on principles and culture.

Marvin Bower, former managing director of McKinsey & Company, calls culture "the way we do things around here" (Deal and Kennedy 1982, 4).

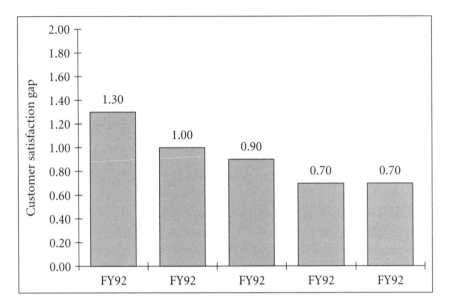

Figure 2.2. Customer satisfaction gap by fiscal year.

It includes ingrained attitudes, values, and beliefs. A positive culture is a prerequisite for lasting change. Programs and procedures come and go, but culture changes very slowly. An organization cannot achieve fundamental changes through "programs of the year" (or month); it must change its underlying culture.

Constraints

A challenge: Undo this incredibly complicated knot, and you will rule Asia. Alexander the Great, pulling out his sword, said, "The rules say, 'undo the knot.' Nowhere do they say I have to untie it." He then slices the knot in half.

We can interpret the Gordian Knot as a set of restrictive paradigms and assumptions that impede creative thought. Alexander the Great's biography shows him repeatedly overcoming or ignoring such paradigms and assumptions. Alexander, in turn, used the mythical hero Hercules as his role model. Hercules, too, often had to find innovative solutions to challenging problems.

Mountaintop's management team realized that it had to look at the organizational culture and its underlying constraints. The plant would have to overcome preconceived ideas, old habits, and restrictive paradigms.

Behavioral Constraints

The principal behavioral constraints included the traditional management-labor relationship and the supervisor's role.

- *Management-Labor Relationship.* The traditional assumption is that management and labor have different, and even conflicting, interests. In the stereotyped relationship, management represents only the interests of the business, while the union cares only about the hourly workers. Management regards the workers as disposable assets: The business hires them when it needs them, and discards them when it doesn't. The union protects its interests with restrictive work rules, and tries to extract as much money as it can. Each party seeks to fortify its interests against the other. The union would, therefore, see radical changes in the organizational culture as potential threats to its interests.

- *The Supervisor's Role.* The traditional assumption about the supervisor's role originates from Taylorism. In its extreme form, Taylorism tells the workers to leave their brains at the factory gate. The supervisor must direct the workers' behavior and give them detailed instructions on how to do their jobs. Management provides the brains, and the workers provide the hands.

The book will discuss Mountaintop's self-directed work teams (SDWTs), which overcame both paradigms. They rely on trust between management and labor, and they discard the traditional supervisor's role. Mountaintop's SDWTs apply the guidance of Armand Feigenbaum (1991, 207): "The most underutilized resource of many companies is the knowledge and skill of employees." Consider a mechanistic organization with 100 professionals and 400 workers that tells the workers to leave their brains at the factory gate. An organic organization that asks its employees to contribute ideas and improve their jobs has five times as many brains as its mechanistic competitor. Tom Peters, coauthor of

In Search of Excellence (Peters and Waterman 1982) says that the frontline worker knows more about the job than anyone else.

Mountaintop decided to involve all employees in every aspect of the business: quality, policies, procedures, productivity improvement, and even customer relationships. Chapter 6 discusses customer contact teams (CCTs) of primarily hourly workers who visit a customer's frontline workers. This is a radical departure from traditional expectations about the hourly worker's role.

Procedural Constraints

Procedural constraints include performance measurements, and the wrong measurements can lead to suboptimization. Goldratt and Cox's *The Goal* (1992) shows what happens when a company uses the wrong measurements to control its manufacturing operations. This book's chapter 9, on synchronous flow manufacturing (SFM), discusses this in greater detail, and Mountaintop had to make changes similar to those in *The Goal.*

Mountaintop used measurements such as equipment and labor efficiencies, and the consequences were similar to those in *The Goal.* To drive efficiencies, the plant tried to run its equipment at maximum load whether the output was useful or not. This practice tied up cash in inventory and did not help the bottom line.

Other dysfunctional measurements compare people and shifts, and create a scarcity or win-lose mentality. They encourage competition instead of cooperation, because everybody wants to look better than everybody else. Comparative measurements forget that the goal is to have the team win, not to have individuals win. Activities focus on winning, or looking good, instead of the overall business. Competitive measurements, like manufacturing efficiencies, can lead to suboptimization.

Logistical Constraints

Goldratt and Cox's *The Goal* (1992) shows the deficiency of the "balanced manufacturing line" model, in which each operation has the same capacity. Much of the semiconductor industry, however, still subscribes to this model. Harris Semiconductor has broken away from this paradigm, and is using measurements that support performance. These are (1) throughput, (2) inventory, and (3) operating expense. The chapter on SFM describes these in detail, but the overall goal is to make money by transforming raw materials into sales.

Throughput, or sales dollars minus raw material costs,* is king because it pays the bills and earns a profit. Goldratt's *The Race* (1996) shows that inventory not only ties up cash, but also affects the company's ability to generate throughput. Inventory can also drive up operating expenses by increasing the need for overtime. Therefore, we want as little inventory as possible. Operating expense is necessary to produce throughput.

To implement SFM, Mountaintop shifted its cost accounting focus from operating expenses and local efficiencies to throughput. This required everyone to change their basic assumptions about true measurements of manufacturing effectiveness. You will, throughout this book, see how Mountaintop broke one self-limiting paradigm after another to transform itself and achieve results.

---

*Throughput is similar to gross margin, but gross margin treats both raw materials and labor as direct costs. The Theory of Constraints (TOC) treats labor as an operating expense, and only raw materials as direct costs. The reason is that, in most factories, labor is really a fixed expense. In practice, labor costs vary only when the factory pays overtime.

CHAPTER 3

Paradigm Busters

William A. Levinson and Jeffrey Lauffer

> *You'll have to find another kingdom; Macedonia isn't going to be big enough for you.*
>
> —King Philip II, to his son Alexander

> *Mountaintop's future is limited only by our imagination and performance.*
>
> —Harris Mountaintop's vision statement

Harris Semiconductor's Mountaintop plant owes much of its phenomenal success to one basic principle: *Success depends on overcoming self-limiting paradigms.*

Joel Barker, president of Infinity Limited, discusses paradigms in his video *Paradigm Pioneers* (1993). Paradigms are frameworks for viewing the world and assessing information. They are helpful when they make it easier to understand new information, but they can also limit our viewpoint. They can blind us to new opportunities or new ways of doing business. Barker also says that today's success can be the seed of tomorrow's failure. This happens if we think "It always worked, so it always will work" and refuse to examine new possibilities.

Obsolete behavioral and procedural models are often more troublesome than competitors. When we limit our mental horizons with preconceptions and prejudices, we automatically limit our opportunities. As soon as we tell ourselves, "We can't do it," we can't do it. General Carl von Clausewitz (1976, book 3, ch. 16) hints at this: "Woe to the government which, relying on half-hearted politics and a shackled military policy, meets a foe who, like the untamed elements, knows no law other than his own power!"

The former government has two problems: its own self-limiting policies and the enemy. History offers several lessons, and most modern examples involve businesses. This book will show frequent examples of how Harris' Mountaintop plant discarded or overcame self-limiting paradigms.

History's First Paradigm Buster

A couple of thousand years ago, the Persian Empire was physically far stronger than Macedonia. King Darius, however, met a foe who knew "no law other than his own power." This is why people in Asia Minor still speak the name "Iskander" with reverent awe.

Alexander the Great's biography shows that he repeatedly overcame and ignored paradigms to achieve results. For example, there was a prophecy that whoever undid the Gordian Knot would rule Asia. The knot, which bound a chariot yoke to a post, was unbelievably complex, and no one could untie it. Alexander looked at the knot, drew his sword, and cut the knot. He realized that the challenge was to undo the knot; nowhere did the rules say he had to unravel it. A paradigm—or preconceived, self-limiting idea—prevented others from applying this very simple solution. Figure 3.1 shows a modern example that trainers use to teach people about paradigms.

At the Battle of the Issus (333 B.C.E.), the Persians followed conventional wisdom. They occupied a strong position and fortified it with palisades. Their position was clearly invulnerable, and any general in his right mind would have realized that an attack would be suicidal. This paradigm would have stopped an ordinary commander.

The palisade sent Alexander a different message. It told him, "The commander who built me has a passive and defensive attitude. He is waiting for you to take the initiative and dictate the battle's terms. Also, you Greeks condition yourselves mentally and physically for face-to-face combat. The Persians behind me are afraid of it, and they're relying on missiles to stop you. Finally, I tether the Persians to their current position. They cannot maneuver without losing my protection."

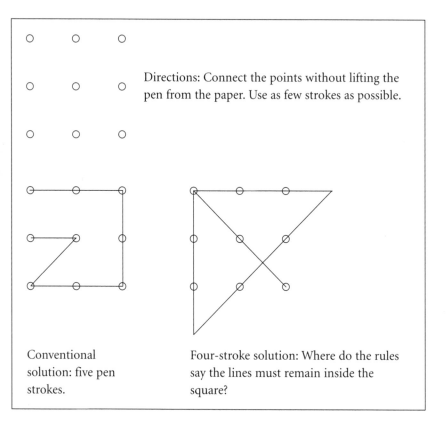

Directions: Connect the points without lifting the pen from the paper. Use as few strokes as possible.

Conventional solution: five pen strokes.

Four-stroke solution: Where do the rules say the lines must remain inside the square?

Figure 3.1. The modern Gordian Knot.

Alexander again defied conventional wisdom, and ordered a frontal assault on the strongest point of the Persian defenses. The Macedonians crossed the Persians' missile zone quickly, so the archery did them little harm. They closed with the Persians, got in their faces, and drove them from the field.

His Role Model

"Hercules the task-solver was to be the god whom Alexander always honored most closely" (Keegan 1987, 18). A Macedonian coin depicts Alexander in Hercules' famous lionskin. To most people, Hercules was a muscleman who beat up various mythological monsters. "He was stronger than the monsters, so what does it prove?" A close examination of Greek mythology, however, shows that Hercules also was a paradigm buster. He often had to outthink the monster, or at least the situation.

Hercules' first Labor was to kill the Nemean Lion, whose hide was invulnerable. Every Greek hunter knew that you had to kill a lion with a spear or arrow. If sharp objects bounced off the lion, the lion was clearly unkillable. Hercules realized that the lion's skin, while impenetrable, had to be flexible, and he killed the lion by strangling it.

Another Labor required him to clean the stables of King Augeus, and no one had cleaned them for 10 years. The king who gave Hercules this delightful task expected to keep the hero shoveling for a long time. Hercules instead diverted two nearby rivers through the stables and washed them clean in a day.

Alexander may have remembered Hercules' alteration of geography when he besieged Tyre. The fortress was on an island and was unapproachable by ships. Alexander built an isthmus from the mainland to the island and brought down the invincible fortress. Alexander's isthmus connects Tyre to Lebanon today.

Rethinking the Role of Engineering

During Mountaintop's transformation, the process engineering department assessed its role in the new environment. The following outline is useful for other departments, too.

1. There is no recipe.

    • A recipe is a self-limiting paradigm. As soon as we say there's one best way to do something, we shut out other options.

2. Back to basics for everyone.

3. Know why your company is in business.

    • A company's goal is to make money now and in the future (Goldratt and Cox 1992). This seems obvious, but many companies have dysfunctional performance measurements that work against it. Chapter 9, on synchronous flow manufacturing, discusses this in detail.

4. Know how your company makes money.

   - A company makes money by selling goods or services to customers.

   - We don't sell SPC charts, design of experiments, engineering projects, operating instructions, equipment, appropriation requests, and so on.

   - This should again be obvious, but it isn't. IBM once rated software programmers on how many lines of code they wrote. "Be careful what you wish, you might get it"—and IBM did. A Microsoft programmer rewrote a 33,000-byte (character) piece of IBM code in 200 bytes (Carroll 1993, 101). Microsoft remembered that customers don't buy software by lines of code, they buy it for what it can do.

     Fechter (1993) cites travel agencies that rate agents on the number of customers they serve every hour. "Travel agents soon learn that meeting customers' needs has less influence on performance evaluation than maintaining a high customers-per-hour average." The travel agency doesn't make money by talking with X customers per hour. It makes money by booking trips and by keeping the customers happy so they'll come back the next time.

5. Know who your customer is—internal as well as external.

   - The book will later discuss internal and external customers. External customers buy our goods or services. Most people in large companies, however, rarely see an external customer. Instead, they provide goods or services to other people in the company. These are internal customers.

   - Manufacturing is engineering's customer.

6. Know what position you play on the team and accept your position. Not everyone can be the quarterback.

   - Again, watch those performance measurements. The performance measurement system is management's magic lamp. As we said: Be careful what you wish, you might (and probably will) get it. Does the measurement system look for superstars, or does it reward team performance?

American popular culture still focuses on individual heroes. Heroes in American police movies (Clint Eastwood's Harry Callahan series, and Bruce Willis' *Die Hard* character, for example) either never call for backup or are cut off from support. Japanese culture, which focuses on groups, has created immensely successful manufacturing organizations.

The cultural differences probably stem from the countries' histories. When pioneers were settling the American frontier, individuals often had to solve their own problems. The nearest neighbor might be miles away, and there was often no police force or nearby army. In contrast, Japanese farmers and fishermen had to work closely together to succeed.

We must understand our country's sociological heritage, its advantages, and its weaknesses. American culture, which stresses individual achievement, produces more Nobel Prize winners than Japan's culture. It produced ground-breaking innovators like Thomas Edison, Henry Ford, and Bill Gates. Japan's culture, however, is very successful in a factory environment.

- "Not everyone can be the quarterback." The sports media is fixated on player statistics, especially quarterback statistics. Sports reporters talk about pass completions, interceptions, and incomplete passes. They forget that a completion requires more than a good throw from the quarterback. A receiver has to catch the ball, and blockers have to keep the defenders from sacking the quarterback before he can throw.

- For example, consider football games like the Pro Bowl and the Blue-Grey game, where each team selects the best players from around the country. The games are less exciting than we might expect, since the players have never worked with each other. They have also practiced under different coaches, each of whom has his own style. It might be interesting to pit one of these all-star teams against a team of merely good players who have worked with each other for a long time.

- As another example, The Penn State Nittany Lions win most of their football games, although they rarely have outstanding quarterbacks. The coach, Joe Paterno, recognizes that the team has to win the game. The coach can't do it himself, and he can't rely on one or two star players, either.

  Chapter 4, on culture as a foundation, will discuss the power of stories and legends. Alexander read Homer's *Iliad*, kept the book under his pillow, and modeled himself after its heroes. Homer's story, therefore, played a role in launching Alexander's conquest of the known world. Virgil's *Aeneid* is similarly behind the success of the Penn State Nittany Lions.

  Paterno read both the *Iliad* and Virgil's *Aeneid*, and quickly noticed the differences between their heroic models (Paterno and Asbell 1989). The Homeric hero was an individual superstar who fought primarily to glorify himself. Homeric ideals stemmed from an era when armies sometimes settled a battle with a duel between their best warriors.

  When Virgil wrote the *Aeneid*, this era was long past. The Roman legion and its predecessor, the Greek phalanx, required close cooperation between team members. While the *Aeneid*'s setting is right after the Trojan War, Virgil probably superimposed contemporary ideals.

  The lesson that Paterno drew from the *Aeneid* was, "You must be a man for others." "Aeneas is the ultimate team man. . . . For Virgil's kind of hero, the score belongs to the team." Like football, manufacturing is a team activity. The Virgilian model is usually more applicable than the Homeric model.

7. Base decisions on what is best for your company, not your particular department.

   - Here come those performance measurements again. Chapters 9 (on synchronous flow manufacturing) and 8 (on total productive maintenance) show what happens when measurements focus on individual departments. The result is suboptimization: good local performance at the expense of the overall system.

8. Get out on the floor and find out what they want by asking/visiting/attending manufacturing team meetings.

   • Tom Peters calls this "management by wandering around."

9. Follow through on what your customer wants. Exceed customer expectations.

10. Believe it, preach it, and live it from the top down.

    • If you aren't a believer, you'll never sell your subordinates on the idea.

    • Management is like riding a horse. Horses are emotionally sensitive animals, and they can tell whether their riders are confident. A rider who isn't confident, or lacks commitment to a goal, can't convince the horse otherwise.

11. Let go, let go, let go of responsibility.

    • Let someone who is closer to the action make the decision for you.

    • The person who is closest to the job usually knows more about it than anyone else. This is usually the frontline production worker who does the job every day (see Figure 3.2).

12. Recognize everyday happenings as opportunities for learning. Learn from problems and mistakes. Use examples of both good and bad behavior to effect cultural changes.

13. Don't lop off heads; make sure you don't single out any one person or department. Spread the learning around. There are plenty of opportunities in every department.

14. Beware of groups that say, "We've *always* worked as a team."

    • Denial is the first step of resistance.

15. Be patient, have more patience, have even more patience.

16. Communicate your expectations and results frequently.

17. Reward success.

    • Chapter 7, on zero scrap, discusses some team recognition and reward programs.

Figure 3.2. Paradigms: Cutting the Gordian Knot.

You will see figures similar to Figure 3.2 throughout the book. They show where Harris' Mountaintop plant overcame a paradigm to achieve outstanding results.

CHAPTER 4

Culture as Foundation

Roger A. Bishop

Culture is an unspoken set of mutual beliefs, values, attitudes, and expectations that guide the behavior of organization members. Courchaine and Williams (1992) refer to the "values and beliefs that influence day-to-day behavior and operations." A strong, positive culture is a requirement for total quality management (TQM) and continuous improvement.

This chapter discusses how attitudes and events have shaped the culture of the Mountaintop workforce. The organization's history plays a key role in shaping its culture, so we will discuss this first. Many events, some good and some bad, can shape an organization's culture. In today's dynamic and turbulent business environment, many organizations are in states of discontinuous change (Hodge and Anthony 1991). If the organization is to survive and prosper, it must adjust all its parts to accommodate change.

Culture is a vital part of any organization, and it pervades every aspect of its operations. Every journey needs a starting point, especially a journey to excellence. This chapter describes the starting point of the Mountaintop journey. The journey has included positives and negatives, but most importantly, it focuses on any organization's most essential asset: people.

Historical Perspective

Mountaintop's history has strongly influenced the plant's cultural evolution. A historical assessment is a logical starting point for any organization that is considering a major cultural change. Mountaintop's story begins in

June 1960, when RCA opened the plant. As part of RCA, Mountaintop supplied solid state components to RCA's vertically integrated consumer electronic product lines.

The opening of this RCA facility was a major event in northeastern Pennsylvania's Wyoming Valley. The community had experienced extreme economic depression because of the demise of the anthracite coal industry in the 1950s. For 20 years after the last major mine closed, the Wyoming Valley's average unemployment rate hovered around 17 percent. The economic situation tore families apart as children left the area to seek employment elsewhere. What remained was an infrastructure in decay and an aging population with little hope of economic prosperity.

Many people have speculated about why RCA would build a plant in such a remote area of the East. An unlimited labor pool and a low wage structure were probably factors. The location also was not far from RCA's divisional and corporate headquarters in Edison, New Jersey, and New York City. The new interstate highway system also made the site attractive. Crestwood Industrial Park is within seven miles of Interstates 80 and 81. What became more apparent was that this RCA facility became the community's hope as a new economic cornerstone.

In 1960, the Mountaintop manufacturing site's mission was to build solid state devices using silicon and germanium technology. The first principal products were silicon power transistors. Later, photo cells, solar cells and other related products joined the product line. Over the next five years (1960–1965), silicon technology—which was growing as a stand-alone business—replaced the old germanium technology. Mountaintop expanded its engineering and services to support the increased activity in the silicon-based market place. By 1966, the plant had virtually doubled its floor space. Further expansion followed as the semiconductor industry grew, and by 1974 Mountaintop's population had reached 3100. In 1977, Mountaintop began producing hybrid electronic products. In 1980, design and advanced engineering were transferred to the plant as the nucleus of a total product development engineering group.

Through this period, the facility's main customers were internal to RCA. "Productivity" meant meeting production and shipment schedules. The workforce was predominately female because the manufacturing process required excellent manual dexterity. This requirement was a perfect fit for unemployed textile and garment workers. The Wyoming Valley's second largest industry, after coal, was textiles. The plant's management structure was traditional, with heavy emphasis on loyalty and dedication. This chapter will discuss the paternalistic structure in detail.

Pressure from Overseas Labor

During the heyday of the RCA Mountaintop plant, solid state electronic technology was fairly simple. Solid state transistors began as replacements for vacuum tubes, but the technology evolved rapidly. In the mid-1970s, overseas competitors entered the solid state electronics field. Most of these competitors were in the Pacific Rim, and Japan was the dominant one.

The final assembly operation, which installs the transistor in a package with external wires, is labor intensive (see Figure 4.1). Most Pacific Rim countries have low labor costs, so U.S. solid state suppliers began to open offshore operations. RCA, which was already an international company, joined others in opening assembly operations in the Far East and opened facilities in Taiwan and Malaysia. RCA chose Taiwan to service specifically the television assembly and component needs, and Malaysia was slated to support the solid state business.

As soon as the Kuala Lumpur, Malaysia operation was ready, RCA moved the assembly operations. The impact on the Mountaintop facility was devastating. Between 1974 and 1985, employment dropped from 3100 to 600. The community perceived that its premier employer was folding in front of its eyes. The facility employees' only hope was that RCA's loyalty toward its employees would prevent it from closing the plant.

Adaptability and Organizational Survival

The choice to move assembly operations offshore was difficult for RCA. The company did not need the capacity of both the U.S and the newly acquired offshore operations. RCA targeted several facilities, including Mountaintop, for closure. However, change is slow in large corporations like RCA. Several factors helped Mountaintop avoid closure. These included adaptability to the changing needs of the business, a stable workforce, and quality of product.

This uncertain period, however, created a culture that relied on negative motivation. From year to year, employees did not know whether the plant would survive. The proverbial carrot was, "You'd better make your

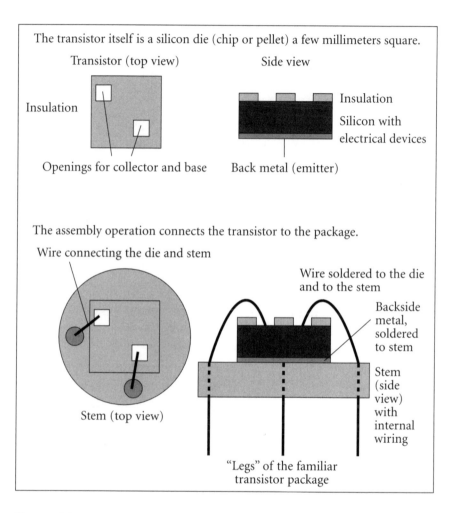

The transistor itself is a silicon die (chip or pellet) a few millimeters square.

Transistor (top view)   Side view

Insulation

Insulation

Silicon with electrical devices

Openings for collector and base   Back metal (emitter)

The assembly operation connects the transistor to the package.

Wire connecting the die and stem

Wire soldered to the die and to the stem

Backside metal, soldered to stem

Stem (top view)

Stem (side view) with internal wiring

"Legs" of the familiar transistor package

Figure 4.1. Transistor assembly operation.

[production] numbers or we won't be here next year." Despite this threatening atmosphere, the operation in Mountaintop not only survived but grew. The solid state industry's evolution into more sophisticated silicon technology drove this growth. RCA Laboratories in Princeton, New Jersey, was one of the research and development leaders in solid state physics. The introduction of microelectronic circuitry was to revolutionize the world we live in. Mountaintop survived because it was able to change from the labor-intense world of manual component assembly to the technology-intensive fabrication of microscopic electronic circuitry.

The Union and Its Influence

When RCA opened its Mountaintop facility, the company had been in business for 41 years. During those 41 years, unions had organized much of RCA's production labor. By 1960, four unions represented nearly 40,000 production workers in the RCA corporate structure. Those four unions were the International Union of Electronic, Electrical, Salaried, Machine and Furniture Workers, AFL-CIO (IUE), International Brotherhood of Electrical Workers (IBEW), International Association of Machinists (IAM), and Teamsters. The IUE and IBEW international organizations had multilocation agreements with RCA. In Mountaintop, it was understood from the outset that the IUE would represent the production and maintenance workers.

IUE Local 177's evolution in Mountaintop differed from other East Coast IUE locals. The greater Wyoming Valley played a key role in the adversarial and sometimes violent birth of the contemporary labor movement. For example, the historical background of *The Molly Maguires* is from neighboring Schuylkill County. The Wyoming Valley figured predominantly in the growth and strength of the United Mine Workers under John Lewis. This historical background suggests that union-management relations at Mountaintop would be volatile and adversarial. The overall labor-management relationship was, however, very stable and cooperative in the Mountaintop plant. The contrast with the Wyoming Valley's history provides valuable insights into the cultural norms of the Mountaintop facility.

Proactive Union Leadership

IUE 177's presidency was very transient until 1976. The union's membership usually replaced the president every two years because of discontent with the most recent contract or because of the challenger's promise for a better tomorrow. This resulted in unstable, short-term leadership. In spite of this, labor relations were good. There has been only one strike in the plant's 36-year history (1970), and one arbitration.

In 1976, a young and inexperienced production operator entered the arena of union leadership. Sally Hooper joined RCA in October 1969, and after experience as a union shop steward, became president in 1976. Her leadership began in the midst of RCA's mass exodus to offshore locations. Hooper faced the challenge of saving jobs for her membership. As a strong analyst, she realized that the union had to align itself with the changing competitive environment. The best way to secure union jobs was to strengthen the business (see Figure 4.2). She took a cooperative and sometimes aggressive stance to achieve this, and she succeeded. She is now nearing her twenty-first year as the president of IUE Local 177. Since her entry into union leadership, she has also assumed a role on the board of IUE International District 1, the overseeing body for all IUE locals on the eastern seaboard.

Consistent, proactive local union leadership promoted the evolution of a very cooperative labor-management relationship. This mutually beneficial arrangement has allowed greater management flexibility and has allowed the business to respond to changing market, technology, and economic conditions. Mutual trust and commitment between management and employees allowed Mountaintop to survive the merger and acquisition cycle that occurred in 1986 and 1988.

Mergers and Acquisitions: The Late 1980s

In 1986, RCA was coming out of one its most difficult business periods. Excessive acquisitions during the 1970s, and frequent changes in executive officers, lead to confusion and ineffective leadership for almost a decade. When the worst was supposedly behind the company, the unexpected happened. General Electric Company, under Jack Welch, recognized RCA's value. GE saw that RCA's parts were worth more than its whole. GE showed particular interest in the RCA subsidiaries of the National Broadcasting Company, RCA Government Systems, and the RCA Service Company. Welch mounted a dialogue with then RCA chief executive officer Thornton Bradshaw. The two men agreed to merge the two companies, but the merger was really an acquisition by GE. GE was then three or four times the size of RCA. As the integration of the two companies began, everyone realized that GE would dominate all business decisions. Within a year of the merger, the consolidation and sale of business units of both RCA and GE began.

GE had been trying diligently to rid itself of it solid state business operations, which did not fit into Jack Welch's strategic plan. He considered the solid state business too capital intensive, and he did not expect it

Figure 4.2. Union-management relationship.

to deliver a 10 percent return on investment. (Ten percent was Welch's minimum criterion for an acceptable business.) He had downsized the GE solid state organization before the GE/RCA merger. The remnants of GE's solid state operations included a few fabrication operations and a West Coast subsidiary, Intersil. With the purchase of RCA, Welch now had a solid state business larger than he had started with. He publicly stated his belief that the RCA solid state organization had excellent leadership and productive manufacturing operations. Nonetheless, he did not want to be in the semiconductor business. Between 1986 and 1988, GE actively marketed the solid state businesses.

Only two years after the RCA/GE merger, Harris Corporation purchased the combined GE solid state businesses. GE Semiconductor, Intersil, and RCA Solid State became the new Harris Semiconductor Sector. The combined annual sales of the GE/RCA solid state conglomerate was then about $600 million. The Harris Semiconductor operation that served the military applications end market was about $300 million. It was Harris Corporation's goal to have a diversified semiconductor business that could serve both military and commercial markets. Harris' purchase of the GE organization accomplished this, since the GE/RCA group focused on consumer products.

Harris Corporation's headquarters is on the eastern coast of Florida, in the community of Melbourne. Harris is among the largest private employers in the state and enjoys a certain degree of favorite son status. Harris is a major player in the electronic systems market, but enjoys little consumer name recognition because of its position as a component and systems supplier. Harris has, for the most part, been a Florida company, with small operations in Illinois, New York, and California. The GE/RCA acquisition was a major expansion in its geographic diversity. This was the single largest purchase in Harris Corporation's history. Harris had experienced much of its growth from external acquisitions. When it bought GE/RCA, Harris' annual sales were almost $3 billion, which made it a *FORTUNE* 50 company.

Cultural Influences

Acquisitions and mergers mix and dilute diverse company cultures. In most companies that grow through acquisition, the culture is often thinner than in those that grow internally. This can be helpful or damaging, and it depends on the company's leadership. Leadership can harm the organization if it undermines strong cultures or damages commitment. It can help by spreading new and useful ways of thinking and by preventing groupthink. Groupthink is actually a disease of strong and cohesive organizations, not weak ones; it is organizational hubris. Everyone begins to think alike, and this means no one is thinking.

The RCA Influence

RCA's policies and actions intentionally developed a parental and thick culture. This was easy, since RCA's name enjoyed respect and recognition in domestic and global marketplaces. RCA was a household word that people associated with influential technologies such as radio, television, and computer systems. RCA always had plenty of tales and folklore of its organizational accomplishments and leadership.

Stories, Legends, and Organizational Behavior. We cannot overemphasize the importance of stories and legends in organizational behavior. Stories reflect the organization's culture, and we can learn a lot about the organization by studying them. Historians can learn a lot about an ancient society's beliefs and values by studying its gods and myths, and organizational scientists can do the same with modern organizations.

Stories also affect the culture, and we can even change the culture by telling the right stories and legends. In ancient times, countries told heroic stories about their historical figures or even assigned them divine status. The Romans said that Aeneas, a semidivine hero of the Trojan War, founded their city. They also numbered Romulus and Remus, sons of the war god Mars, among the nation's founders. Roman standard bearers wore lion skins to emulate the legendary Hercules, who slew the Nemean Lion. Meanwhile, portraits of Hercules on Macedonian coins bore amazing resemblances to Alexander the Great. Julius Caesar deliberately created legends that served his needs, and then the Romans began deifying their late emperors.

Modern organizations often develop larger-than-life tales about their founders. Thomas Edison and Charles Steinmetz are legends at General Electric, while Thomas Watson Sr. is a legendary figure at IBM. Tom Peters and Nancy Austin (1985, 328–329) cite Alan Wilkins, who said, "some kinds of stories are powerful ways to motivate, teach and spread enthusiasm, loyalty and commitment; others served an equally powerful purpose: to perpetuate cynicism, distrust, and disbelief." RCA's history promoted loyalty, trust, and commitment.

RCA's Culture and Employee Commitment. RCA's labor relations were never as volatile and controversial as General Electric's. There was, however, a definite separation between labor and management, especially at the manufacturing locations. RCA's policy had been to keep nonrepresented employees' benefits and wages ahead of union employees'. This created two subcultures, which RCA called salaried employees and hourly employees. The difference went beyond the method of pay distribution, and extended to expectations about the employees' roles. It was easy to see an "us" and "them" culture in any of the RCA facilities. Although the company environment strongly reinforced this subculture, both groups were still intensely loyal to the parent organization. While the company expected hourly workers to take direction and not think for themselves, it also cared for their welfare.

The following example shows why RCA enjoyed its employees' loyalty and commitment. In June 1972, Hurricane Agnes struck the East Coast and flooded Pennsylvania with torrential rains. The Susquehanna River overflowed its banks, which put many low-lying regions under water. Many employees at the Mountaintop plant lost all their worldly possessions to Agnes' flood waters. As they tried to rebuild their lives and homes, money became a major issue. Most had insurance, but they needed to pay for clothing and shelter until the insurance payments came through. RCA offered every employee who lost his or her home a cash, interest-free loan of $1000 for this purpose. The only document to secure the loan was the employee's signature on a note. A true financial manager would balk at loaning hundreds of employees cash with no collateral. RCA ignored the prospects of possible defaults and loaned the money anyway. The employees repaid every single loan, proving that loyalty and commitment are better surety than collateral.

This was only one example of the close relationship between the company and its employees' personal lives. Employees could purchase consumer electronic products from the RCA "Family Store" via payroll deduction. The company provided fully paid benefits to members of the "RCA Family." The company had an annual family picnic, employee Christmas party, children's Christmas party, Thanksgiving dinner, and so on. Each event reinforced the idea that the employee was a valuable and loyal member of the closely knit organizational family.

In Mountaintop, this cultural norm was easy for new employees to accept. The Wyoming Valley's predominant ethnic makeup of eastern European cultures, with their strong needs for family and community ties, supports this company culture. The workplace became an extension of the family. It wasn't enough for new employees to do their jobs; they had to fit in socially with the work unit. It was difficult, if not impossible, to do your job efficiently and not be accepted by the work unit.

Uncertainty and Insecurity. Working the Mountaintop plant during RCA ownership was an uncertain proposition. This was especially true after the dramatic employment reductions during the 1970s. Strategic plans were not in place, and layoffs were common. Employees were not considered seasoned veterans unless they had been laid off at least once, and this uncertainty lead to extreme insecurity. Employees took second jobs in case the plant closed. Despite this trauma, however, loyalty and dedication persisted. The Mountaintop facility's voluntary termination rate has never exceeded 1 percent, except for retirements. This is very low—especially for the semiconductor industry, whose annual attrition rates range from 10 to 16 percent. Mountaintop's fortunes were appearing to improve, and then the merger hit.

A merger is traumatic for most employees in any company, but it was devastating for RCA's employees. First, there was no warning or signal that a merger was happening. Many RCA employees also perceived a betrayal by RCA's corporate management and by the clandestine events leading up to the merger. In the two to three years before the GE acquisition, RCA struggled financially. Interest payments on long-term debt consumed most of the available cash. Management responded with an all-out effort to regain financial control of the corporation's 270 locations and businesses spanning hundreds of products. The company's leaders asked management to restore the corporation to its previous status, and management passed this message to subordinates. Three years of effort produced major gains, and the company had apparently turned the corner to success. The employees had achieved what corporate leadership had asked. What was the reward? Betrayal—or so thought many of the RCA employees who heard about the merger for the first time on the six o'clock news. For two years or so following the merger, members of the RCA family watched as the new owners dismantled the corporation. When it was over, the corporation known as RCA no longer existed.

The GE Influence

In Mountaintop, the merger announcement's impact was the same as at other RCA locations. "Shock" best describes the reaction by the general Mountaintop plant population. The reaction from local labor officials was, however, even more dramatic. General Electric did not have a good reputation in labor relations. Organized labor did not see the company that coined the phrase *Boulwarism* as positive. Boulwarism means that management puts its "best and final" offer on the table at the beginning of union negotiations. The message to the union is, "Take it or leave it." Questions about GE's dedication to the semiconductor business worsened the atmosphere. Meanwhile, GE directed the site's management to increase output beyond order requirements. This resulted in a large finished goods inventory. After the plant built the inventory, it began to receive visitors whom management was told to call "customers." It took the employees in the plant about 30 seconds to realize that these customers were prospective buyers. They came from Italy, Japan, Illinois, and California. Each looked at the equipment, reviewed the labor force statistics, and examined the plant financial statements. The Mountaintop organization, however, kept functioning under the same basic policies that were established under RCA. GE actually never changed the sign in front of the building. This small signal made employees feel even less secure, and daily rumors circulated about either a sale or closure.

During GE's ownership, the RCA culture eroded under the strains of uncertainty. There were certain GE corporate mandates but, as good conquerors do, they kept many RCA managers. GE was willing to let these managers run the business until its final disposition could occur. This was when I saw the effects of organizational culture and its breakdown. Employees went through the motions, but they lacked purpose. Their paycheck still used RCA check stock, and the name on the building was still the same. However, they didn't know who they worked for. We were officially the RCA Solid State division, a wholly owned subsidiary of General Electric. This answer, however, provided no relief from the anxiety created by the perceived betrayal of some faceless entity.

General Electric's failure to integrate Mountaintop into its own culture further confused the employees and made them distrust the corporate institution. Mountaintop's management escaped the employees' discontent because employees still saw them as part of RCA, not GE. This perception, and management's freedom to operate under the old policies, allowed the business to keep meeting its customers' demands.

The Harris Influence

The date was December 1, 1988. For five to six months, confidential meetings had taken place between General Electric and Harris Corporation. GE was eager to rid itself of the semiconductor businesses. Harris was, in contrast, anxious to diversify. It was an opportunity for both companies to realize their short-term goals.

Harris sent several financial and senior management teams to Mountaintop for a pre-purchase assessment. Among them was the vice president of human resources for Harris Semiconductor. His agenda was to assess the labor relations climate and employee attitudes toward a purchase so soon after the GE/RCA merger.

What was the climate at this point? As discussed earlier, Mountaintop employees saw GE as a threat and as an undesirable owner. The employees blamed GE for the demise of RCA Corporation. The feelings were so strong that employees in some former RCA locations held mock funerals for "Nipper," the RCA mascot (the dog with the phonograph). In Mountaintop, the local union had ordered and sold RCA memorial sweatshirts. The employees realized that GE would eventually find a buyer or, in the worst case, close the operation. Employees were also afraid that GE might sell the plant to a Japanese firm. From organized labor's perspective, this would be a fate worse than death. When Harris and GE announced an "agreement in kind," there was visible relief in Mountaintop. A series of questions, however, followed. The most frequent question was, "Who and what is Harris Corporation?"

As a major supplier of military electronic systems and components, Harris was not a household word. Most employees in RCA's electronic components businesses did not see Harris as a direct competitor. Harris' acquisition of the GE solid state conglomerate, however, made the company a major competitor in the commercial semiconductor marketplace. The purchase of the GE solid state group was Harris' biggest single investment since the company's birth in 1895. The Mountaintop employees worked their networks to glean every piece of information possible on

Harris Corporation and particularly the semiconductor sector. Meanwhile, the international organization of the IUE used its research department to provide the local union officers information on Harris' structure, leadership, economic position, and—most importantly—its union affiliations. What they found was not reassuring. Harris, from the days of Harris Linotype, had a reputation of being a tough negotiator. Harris was also a mostly nonunion company. Mountaintop's IUE Local 177 now found itself as one of only three unionized Harris plants. Would Harris recognize the union under the terms of the sales agreement?

Harris' Assimilation of the GE/RCA Businesses. Mountaintop's employees didn't know that Harris' CEO, John Hartley, wanted to smoothly integrate the GE/RCA businesses into Harris. Harris was striving for positive integration when it first opened discussions with union leaders. Harris told the union representatives that it would recognize the unions at former GE/RCA locations. The company was also willing to recognize employees' prior service credits as Harris service. This was the first of several policy decisions that had a positive impact on the Mountaintop employees. It lessened some concern toward Harris' intentions.

The first step in the labor relations process occurred with an interim contact agreement in Washington, D.C., with IUE president William Bywater. That agreement bridged the existing contract with GE/RCA until expiration. It included full acceptance of the existing union, wage structure, and total one-to-one benefits exchange. During this discussion, international officers briefed the local union on the options Harris could have pursued. The integration process therefore began with Harris' show of good faith.

In the Mountaintop plant, word reached the workforce of Harris' acceptance of the RCA and GE employees. About a month after the sale, a human resources representative presented an overview of the Harris benefits programs. However, the employees still had limited information and exposure to Harris' management and culture. As the integration process progressed, several issues had a direct bearing on the cultural change process. The first issue related to the Harris infrastructure. The acquisition of a large organization like RCA/GE/Intersil tested Harris' functional support systems to their limits. This prompted an immediate internal focus to lessen the strain's impact on the operational aspects of the semiconductor business. The company had to pay a lot of attention to managing the infrastructure and consolidating redundant functions. These distractions forced management's attention away from customer and market issues. This was also true in the product lines located in Mountaintop. This caused customer dissatisfaction and lower productivity and profitability. Two years after the acquisition, the Mountaintop product lines were in the red. By 1990, we were losing an intolerable $20 million per year. A major reduction in labor and operating expense occurred between 1990 and 1991. The company laid off 45 percent of the salaried and 25 percent of the hourly workforce. During this time, Harris' sector management introduced the "Quality by Design" concept. This was to become the forerunner of self-directed work teams, which the following chapters will cover in detail.

These dramatic changes, and the employees' desire to stay in business, helped the Mountaintop culture evolve and mature. The "Quality by Design" process, then the SDWT concept, empowered the employees and made them feel like part of the business. The culture changed from parental reliance to individual and organizational accountability. Harris' senior management allowed Mountaintop the flexibility to encourage employees to participate fully in their own recovery. The employees' tone changed from, "Why don't you [management] do this?" to "Why don't we do this?" This new freedom encouraged workers to involve themselves in a new cooperative process. The production workers could no longer lean on supervisors, but they also faced fewer bureaucratic obstacles. They began to identify and implement cost reductions and efficiency improvements, and the empowerment was successful.

Mountaintop's Culture, Empowerment, and Commitment

> *Say 'we,' 'us,' and 'ours' when you're talking,*
> *instead of 'you fellows' and 'I.'*
>
> —Rudyard Kipling, *Norman and Saxon*
> (or "Managers and the Workforce")

This success has continued and has been the foundation for the continued evolution of the self-directed work teams. More importantly, it has established a Mountaintop culture that has eroded the "us" versus "them" philosophy of the past. The new culture is indeed "we," "us," and "ours." It has also provided employees with positive motivation and provided a feeling of accomplishment and pride.

This cultural evolution had its price. For some employees, empowerment meant abdication by management. There was confusion at the onset. One day, employees received direction; the next day, they didn't. Employees had to discard ideas like "This is my job, and that isn't." Many employees had lost close associates to the downsizings.

The biggest change was the elimination of the production supervisor position. Ninety percent of the former supervisors received new assignments, but some saw those moves as demotions. In reality, 73 percent of the supervisors received lateral transfers into team supporting roles (that is, technicians and production schedulers). Nineteen percent took a voluntary layoff package, mostly because of philosophical differences in the organizational directive. The remaining 8 percent could not function in the new culture and lacked transferable skills. The company had to discharge them. Their reactions were mixed and, at times, bitter. Again, the 28 years of parental pretense was not going to disappear overnight. These feelings, however, were deeply entrenched in less than 15 percent of the population. The remaining 85 percent saw the changes as an opportunity to correct the errors of many years and, by doing so, to ensure the survival of their positions.

Today the culture grows and strengthens with employees' development and their awareness of what their contributions have done for the organization. We cannot erase 36 years of history, but we can view them as learning experiences instead of inevitable results. We can learn from the past, but we need not be its prisoner. Mountaintop had to adopt the new cultural base to survive, but it is now the foundation for continuous improvement.

Theory and Practice

When Mountaintop began its cultural change process, management looked for a road map. After many months and dollars it found a simple answer: There is no road map. The dynamics of group development are critical in change management and especially in cultural norms. Some organizational theorists believe that the formation of groups (in Mountaintop's case, self-directed work teams) follows the sequential stage theories. These were espoused by W. Bennis and H. Shepard in 1956 and by contemporary authors like B. W. Tuckman and M. A. C. Jensen in 1977.

Sequential Stage Theory

Tuckman and Jensen (1977) call these stages forming, storming, norming, and performing (see Table 4.1). In the *forming* stage, the group is establishing a climate and getting to know one another. The second stage, *storming*, is the sorting out of expectations and joining forces stage. The third stage, *norming*, is when group agrees on norms and guidelines. The fourth and last stage, *performing*, is when the group begins to perform as a team. As Tuckman and Jensen see it, a performing group can function effectively. Some organizations, however, mistakenly view the performing stage as the conclusion of the group's development. Groups, like organizations and individuals, must improve and develop to meet the challenges of a changing environment.

Recurring Phase Theory

Another group of theorists, however, believes that group development evolves in recurring phases. Behaviorists such as R. E. Bales (1950) and W. Schultz (1966) support this theory. The basic premise is that groups are in a continual state of development. Although they go through three basic stages, their growth does not end at the performing stage. Instead, the group will continually experience the process as change occurs within or outside itself (see Table 4.1). R. E. Bales (1950) refers to the first three stages as *orientation, evaluation,* and *control.* In the orientation phase, the group gathers information and clarifies tasks. Evaluation means assessing the available information, and control means deciding what to do. Bales believes that problem-solving groups continually face two similar but distinct concerns. Task-oriented concerns relate to efforts to accomplish the group task, while socioemotional concerns refer to the relationships among members. Both concerns operate continually. The group's attention to one of these may produce strain on the other (Napier and Gershenfeld 1993, 491).

Table 4.1. Group development theories.

Theorist Sequential stage theories	Development stage 1 (the feeling out and establishing climate phase)	Development stage 2 (the setting expectations, sorting out and joining forces phase)	Development stage 3 (development of norms and guidelines established phase)	Development stage 4 (the productive phase of the group)
B. W. Tuckman and M. A. C. Jensen (1977)	Forming	Storming	Norming	Performing
P. Hare and D. Naveh (1984)	Latent	Adaptation	Integration	Goal attainment
W. Bennis and H. Shepard (1956)	Dependence	Interdependence	Focused work	Productivity
B. A. Fisher (1970)	Orientation	Conflict	Emergence	Reinforcement
Reoccurring phases theories	[All stages taking place at the same time]			
R. E. Bales (1950)	Orientation	Evaluation	Control	
W. Schultz (1966)	Inclusion	Control	Affection	

Mountaintop's Experience

When we began our group development activity in Mountaintop, we used the sequential stage model. As we prepared and educated our workforce for empowerment, we considered the effect of the group development process. We expected the teams to remain functional when they completed the process. We found, however, that the empowerment process often stalled at successive phases. *Empowerment stages* refer to the level at which management lets employees perform outside their normal task responsibilities. In Mountaintop, we began by placing employees in teams organized by natural work groups.

Management first empowered employees to schedule their own lunches and breaks as long as they maintained production. There was no longer a first line supervisor to tell them when to take their free time. The only monitoring came from coworkers, but they did not expect confrontation from peers. Conflicts arose as they tried to agree on break schedules. It was an apparently simple task, but there was no longer a supervisor to act as referee. (When supervisors had arbitrated such disagreements, however, both employees were rarely happy.) Team members were unwilling to change their behavior because, at first, they saw no need to do so. However, experience and training helped them learn to handle these situations. We then assumed that they were ready to apply this model to future issues.

We were wrong! The next level of empowerment was responsibility for vacation scheduling, and the problem repeated itself. New information, processes, and boundaries resulted in new conflicts between task and socioemotional concerns. The formerly important activities of scheduling lunches and breaks now gave way to vacation scheduling. This pattern recurred as the natural evolution of team empowerment progressed. This experience taught us that teams never stop at the performing stage. Instead, they repeat the team development process as they face each new problem. These teams evolve at their own pace. We naively assumed that all teams would progress at the same rate if we provided the same education, training, and guidance. When management recognized that each team evolves at its own pace and that team development is ongoing, it was in a better position to support the teams' needs.

Strategic Beliefs in the Role of Insignificance

I discuss self-directed work teams with organizations that are benchmarking our organization, with associates, and with my students at the university. These discussions usually turn to the traditional rank-and-file employee. This is not surprising, but the lack of questions on management change is. The Mountaintop experience has shown that the change process affects the management culture, too. Why is the management culture important? W. Scott and D. Hart (1990) define the modern organization as "managerial systems, using universal behavioral techniques and communication technologies, to integrate individuals and groups into mutually reinforcing, cooperative relationships." At the center of their hypothesis are the "strategic beliefs in the role of insignificance."

Scott and Hart divide their theory into positive and negative beliefs (see Table 4.2). Managers usually subscribe to the positive beliefs:

Table 4.2. Positive and negative beliefs (Scott and Hart 1990).

Positive Beliefs	Negative Beliefs
Human nature is malleable and can be shaped to suit the organization's needs.	Innate humane motives do not exist, and therefore innately necessary human needs do not exist.
Intelligence is necessary for rational production and consumption. Therefore, its cultivation is important to the modern organization.	Intellect is a dysfunctional form of personal fulfillment for most people, since it tends to separate people for the collective enterprise of the modern organization.
The modern job is the chief source of personal satisfaction for people committed to the collective enterprise of the modern organization.	Work is an undesirable means of individual expression in the modern organization because it results in variability rather than uniformity.

Human beings are naturally willing to do what is best for the organization. To do so, however, they need intellectual preparation to deal with the modern organization. Also, a job is a person's chief source of personal satisfaction.

Workers in traditional organizations often subscribe to the negative beliefs. Under this system, people do not need motivation, especially not by organizational requirements. This viewpoint also defines intellect as a dysfunctional form of personal fulfillment. Intellect separates people into classes and makes them unequal. Individual expression in the modern work environment produces variation instead of uniformity. It prevents the application of basic theories of equity.

Under Scott and Hart's theory, how can an organization evoke change? Real people, whether management or rank-and-file, share both the positive and negative beliefs. The difference is in how each person applies them to his or her situation. The person's need (desire) to create change also will have an effect. This theory applies to a person's willingness to change for the good of the organization. Mountaintop's experience shows that change does not happen unless there is a need. Mountaintop had to change to survive, and each employee wanted to protect his or her livelihood. Therefore, the positive and negative belief systems found a middle ground because of the organizational need to survive.

Need as a prerequisite for change presents a unique challenge for profitable businesses. They must accept organizational and cultural change to prepare for the future, but there is no visible, immediate need to do so. Such organizations must teach their workforces about the change process. This can be an exceptionally powerful tool in helping employees find that middle ground.

The most illustrative story of the need for change is that of the frog in a pot. If you throw a frog into boiling water, it will jump out immediately because it recognizes the need for change. If you put the frog in warm water and heat the water slowly, the frog will stay there and cook because it never recognizes the need for change. A World War II political cartoon also conveys the idea. It showed Uncle Sam lying face down and freezing to death in a snowstorm labeled "Nazism." The caption said, "Leave me alone, it's so peaceful." The Pearl Harbor attack was the United States' wake-up call.

Niccolò Machiavelli (1965, 20) recognized that organizations rarely change without immediate, compelling reasons. He compared statecraft to consumption (tuberculosis) that, in its early stages, is difficult to diagnose but easy to cure. Only a wise physician (or prince) can diagnose the problem in its early stages. When the disease gets bad enough that it is easy to recognize, it is difficult to cure. That is, an ordinary organization needs a wake-up call before it acts to secure its future. An extraordinary, proactive organization doesn't, but this level of proactivity requires awareness throughout the organization.

Culture's Effect on the Change Process

The book will later examine the change process in Mountaintop. We will now look at the existing culture's influence on the first attempts to change the organizational norms. This happened before our full involvement in self-directed work teams. The program was called Quality by Design. In 1990, the Harris Semiconductor Sector management was looking for a way to change how the business units operated. Employee empowerment was clearly a key to this change. Management sought external help in designing a change model, then looked for a place to try it. Mountaintop was one of those sites.

Initial Failure and Its Lessons

We began by selecting three departments in the Mountaintop manufacturing organization. Two were mature operations, and the other was a new production operation. The 100 members of those organizations each received 24 hours of team training. At this point, there were first line supervisors, who attended the training with the production personnel. We felt that the beta site concept would let us assess our methods and adjust them for the rest of the organization. (The term *beta site* often refers to an experimental or trial location.)

The process, however, did not meet our expectations. The rest of the employees saw those in the teaming process as "special."* They perceived those in the beta groups as having privileges that placed them outside the rest of the organization. The other employees treated the beta teams differently and ostracized them for their involvement in this experiment. The first line supervisors also perceived the process as a threat to their traditional role. Instead of supporting the program, the supervisors perpetuated a business-as-usual attitude among their subordinates. The supervisors continued to give direction, so the work teams did not learn to manage their own activities. This was especially true in the more mature work groups, who felt that the old system worked and there was no need to change. In the new operation, the employees were more receptive. The supervisor's wide span of control helped make the employees participate, since the supervisor didn't have enough time to provide close direction. The employees in this group were also self-starters, who wanted to take greater responsibilities. The effort failed, however, because of the problems with the other groups. The site scrapped the program after six months.

What did we learn from this first attempt? In a thick culture, everyone must understand the need for change. The program must be universal, not piecemeal. It must use a standard and comprehensive set of boundary

---

*Readers may also recognize the "Hawthorne Effect." An experiment with lighting improved productivity, not because the lighting affected the job, but because the employees who participated were responding to the extra attention.

conditions. The management structure must support the change process, or it will fail. Employee will be suspicious if they see management singling out individual groups for special attention. This is especially true in a union culture, where equality is a key value. Employee suspicion and distrust will undermine the change process. Most importantly, change and acceptance of change occurs slowly. Also, the bottom-line results are not immediately measurable. Few quality programs yield instant profitability, expense reduction, or reduced cycle time. Too many managers expect to adopt a "program of the year" (or month, depending on their attention span and patience) and get instant quality and profits. These will happen, but they often take longer to achieve than management would like. For those who try this cultural change, there is a price: lost time, training costs, and support systems and materials. For us, this small investment produced a huge return.

Mountaintop's experience was a perfect demonstration of Stephen Covey's (1991) "Law of the Farm." The farm teaches that there are no easy, instant success formulas. We must plow and plant in the spring, and tend the crops in the summer, before we can harvest in the fall. We cannot sow today and expect to reap tomorrow, although this is exactly the attitude behind instant success formulas.

The Process of Change

Changes in organizational culture and norms affect internal and external relationships. This section discusses the empowerment model from a corporate, sector, and plant perspective. It also discusses the impact of these changes on labor relations, customers, and the community.

Harris Corporate and Semiconductor Sector management supported the implementation of the empowerment model at Mountaintop. The support was passive, since Mountaintop's management team knew more about the plant's culture. It was better to let the managers who knew the culture play the active role. The Semiconductor Sector had already succeeded with self-directed work teams in the Palm Bay, Florida location. The next step was to expand that success to other locations through the Quality by Design process. Mountaintop's first attempt to implement Quality By Design hadn't worked, but the management team was willing to take the challenge again. There was an excellent director of operations in Mountaintop to make this happen. Raymond Ford was a natural proponent of employee involvement, and he quickly ordered a new direction for Mountaintop. His vision was straightforward: less bureaucracy and more employee participation. He addressed the entire Mountaintop population, and presented his vision of Mountaintop and their role in that vision. He made it very clear that this decision was not subject to debate. He expected 100 percent participation, and there would be no tolerance for anyone who obstructed the changes.

Ford's approach to change management used elements of the three change models: empirical-rational, normative-reeducative, and force-coercion. The empirical-rational approach persuades people that the change is good for them. Mountaintop had to change to survive and compete in the 1990s. The normative-reeducative model changes group norms, personal attitudes and beliefs, and the culture itself. The organization internalizes and assimilates the changes; they become part of "the way we do business here." This was, of course, what the plant wanted to accomplish. Force-coercion is the least effective method, at least when used by itself. Some people will actively embrace the change and become change agents. Others will go along with the changes, but a few will try to obstruct them. One must sometimes tell people to "lead, follow, or get out of the way."

The time was right for the Mountaintop culture. Traditional approaches to the site's problems hadn't worked, and we were on the verge of closure. We now had leadership that offered a new but unproven solution. However, the workforce needed direction and confidence to accomplish the goals. It was that belief in themselves and the organization's culture that allowed Mountaintop's employees to accept and adopt the cultural change process.

Labor Relations and Change. Organized labor has written many articles on employee involvement. Unionized supporters see employee involvement as an opportunity for union members to control their lives and participate in the organization's rewards. Its opponents see it as a ploy to extract more work from the workers for less pay. They also see it as a way to erode the basic union structure by the formation of company unions.

The new Quality by Design teams raised many questions in Mountaintop. However, there was open communication between the IUE and local management. The union voiced its concerns, and management offered a forum to discuss the issues openly. The officials and members of the IUE Local 177 were initially skeptical of the process. Part of the skepticism came from the fact that they were dealing with new managers from the Harris sector operations. The plant's history was, however, an even bigger cause. For 28 years, the plant had operated in a highly controlled, traditional environment. Management did not ask employees to participate, and expected them to leave their brains at the front door. The Quality by Design process offered the employees far more freedom and asked them to take far more responsibility.

The philosophical support of the IUE International helped get Local 177's involvement. In early 1990, Edward Fire—secretary/treasurer of the IUE International—and members of the IUE District One executive management met with officers of Local 177 and several members of Harris management. During this brief meeting, Fire described the successes of employee involvement teams at General Motors. The IUE had worked with the UAW to set up these teams at GM. He spoke of the need for unions to work with management to make the business competitive. He also described how unions could cooperate within the boundaries of the collective bargaining agreement. The partnership would prepare the company to compete in a global economy. Through Fire's words of encouragement and the efforts of Local 177 president, Sally Hooper, the cooperative participation in the cultural redesign process began (Bishop and Murphy 1993).

Since that meeting, the officers of IUE 177 have supported the cultural transition. The nearly complete absence of employee grievances attests to the transition's success. It took less than three weeks to negotiate the last three union contracts with a win-win bargaining strategy. In the most recent contract, the union membership ratified a wage package that included a profitability-dependent wage component. Mountaintop could not have achieved its cultural changes without the union's support.

Customers and Community. Before Mountaintop began its cultural change, its relationships with customers were, at best, strained. Issues regarding delivery and quality problems were pervasive, and the plant's market share was eroding. Harris set up a GAP index to measure customer satisfaction. Customers rate Harris on several quality elements, then also rate the elements' importance. The GAP index is the difference between the plant's performance (as perceived by the customer on a 1 to 10 rating scale) and the element's importance. A high GAP number means performance is less than the customer's value assigned to that element, while a low number means performance exceeds it. Once a customer has reached a GAP number higher than 2.0 you have probably lost them as a customer. We contacted several of our major customers every month and asked them to rate us on overall performance and certain elements. Customers prioritized performance factors, such as quality and on-time delivery, and rated the plant's performance against those metrics. Our target was a GAP rating of 1.0 or less, but our organization's first overall GAP score was 1.3. Our GAP scores are now consistently between 0.7 and 0.9.

The key point is the direct correlation between organizational behavior, organizational culture, and customer satisfaction. This has been particularly true in Mountaintop. Bringing the employee closer to the customer has helped us understand and fill customers' needs. Many of the problem-solving techniques taught in our team member training apply to customer issues.

The community barely noticed the sale of Mountaintop to Harris. The major publicity about the Harris acquisition occurred during the massive workforce reductions in 1990. Until then, the community believed the plant was an RCA facility. The Mountaintop plant population, in 1990, was around 500. The rumors of closure were so strong that the Chamber of Commerce was calling to see what it could do to save Mountaintop. Today, the Mountaintop facility enjoys the status of a premier and progressive employer. We receive two to three requests for plant tours every week. Our site managers are in demand for speaking engagements for local organizations. Our current $250 million expansion program is the largest single investment in northeastern Pennsylvania history. There are 75 applicants for every production job, although the local unemployment rate is less than 7 percent. This positive publicity, and a steady record of successes, reinforces the new Mountaintop culture. As the saying goes, "nothing is better for success than success itself."

The Impact of Cultural Change: A Human Resource Perspective

I have been in the human resource profession for more than 23 years. I had some initial misgivings about the cultural change process, but you are reading the words of a believer. Mountaintop's approach may or may not work in other organizations, since the plant's history and the local community's culture were factors. It is dangerous to apply a blanket approach to an organization, and we are not offering our experience as a canned, prepackaged approach. The change agent must instead look at his or her organization and tailor the approach accordingly. This chapter shows that positive cultural change is achievable, and it offers some lessons.

Managers must be willing to break paradigms, and they must show commitment to the change process. This is particularly true in human resources. The empowerment model deals with the employee freedoms in self-management. Instrumental in employee policy development has been the human resources activity. Will your human resource function see empowerment as an opportunity or a threat? The answer to that question may determine whether you are ready for cultural change. The human resource activity does not have to lead the change; it is better if it doesn't. The change process must be an organizational imperative that involves all levels and functions.

We cannot overemphasize the importance of management commitment to the change process. Managing an organization is like riding a horse, and the challenge of jumping an obstacle is an example. The rider must be confident that the horse is going to clear the obstacle. The horse will sense a rider's lack of confidence or commitment and will balk. If the rider has unequivocally decided "we *are* going over this fence," the horse will do it. In business, the fence represents self-limiting paradigms and obsolete ways of doing business. Organizational survival may depend on clearing this fence.

The Human Resource Department's New Role

The Mountaintop human resources function has redefined its organizational role as a result of employee empowerment. Its role as a policy enforcement branch has diminished. The expectation is for human resources to function as a supporting member of the teams. This places human resources in a customer-client relationship. The human resource activity's success depends on the accomplishment of the organizational goals. Historically, this was not always the case. We now ask employees to participate in their conflict resolutions. The term *employee relations* has become *facilitation assistance.* In labor relations, the communication channels are much broader. The traditional labor relations management role is now shared with operational management and members of the self-directed work teams. Even the union shop stewards' role has changed to meet the new organizational norms.

When we survey the Mountaintop employees, we ask, "Are things better or worse since the last survey?" Their response has consistently been "Better." We have asked whether they want to return to the traditional organizational structure. The response to that question has been an overwhelming "No!"

Does this process ensure harmony and a total quality management process? The answer is no. Only an environment of continuous improvement that involves all employees will ensure total quality. Has Mountaintop reached its goal for organizational harmony? Again, the answer is no. Organizational harmony is not a destination, but a never-ending journey of continuous improvement.

CHAPTER 5

Teaming to Win

Martin L. Wentz

Chapter 4, on culture as foundation, discusses the Mountaintop culture and the need for change. Mountaintop's shift to a self-directed work team (SDWT) culture was among the plant's most important changes. SDWTs, or autonomous work groups, have played a key role in Mountaintop's competitiveness. This chapter will show how Mountaintop started them and made them work.

In 1991, the organization was not achieving the necessary results. The plant manager and his staff used the Goldratt thinking process to study the situation at Harris Mountaintop in 1991 (see Figure 5.1).

The staff realized that the organization itself had to change. The management hierarchy (see Figure 5.2), which carried over from the RCA days, was not the correct system to lead the plant into the 1990s.

This structure separated the organization into many disparate groups that performed similar functions. It also separated the manufacturing support functions: equipment support, process technicians, and maintenance technicians. There were redundant engineering and manufacturing organizations for the bipolar wafer fabrication and the metal oxide substrate (MOS) wafer fabrication. Everyone was working hard, but the plant was losing money. When studying an organization, we must remember: "Don't confuse motion with progress."

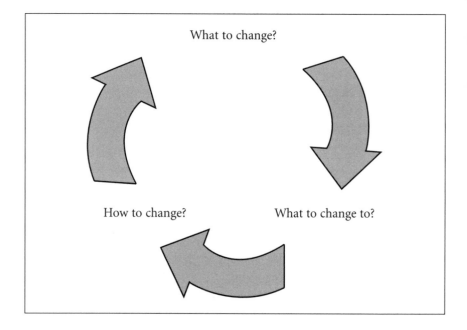

What to change?

How to change?   What to change to?

Figure 5.1. The Goldratt thinking process.

Structure

An organization's structure has a strong influence on its performance. Mechanistic systems rely heavily on rules, procedures, and hierarchies. Organic systems encourage their members to apply judgment and initiative. Mechanistic systems work well in a static and well-defined competitive environment, but this doesn't describe the late twentieth century. Organic systems are agile and adaptive, and can respond to a turbulent, dynamic environment. In practice, organizations have both mechanistic and organic characteristics. In 1991, Mountaintop's structure was primarily mechanistic.

Readers of Tom Peters' books will immediately recognize a problem with Figure 5.2: the presence of organizational barriers. W. Edwards Deming says to break down barriers within organizations. Peters (1988) extends this guidance by urging readers to make their organizations porous to customers. He also says that the person who does the job knows more about it than anyone else. Stephen R. Covey (1991) says that

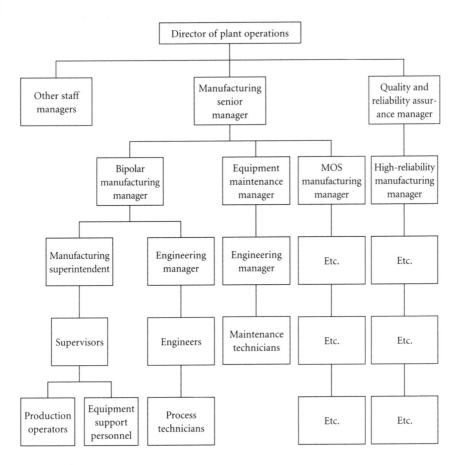

Figure 5.2. Mountaintop organization before SDWTs, 1991 (simplified).

workers who understand and internalize the organization's vision, principles, and sense of purpose can accept enormous responsibility.

The self-directed work team combines these ideas. It encourages workers to apply judgment and initiative while taking direction from the organization's vision and principles. It promotes organizational porosity by bringing together people with different skills and perspectives. We said that quality programs and activities should support each other. The same lesson applies to principles for organizational leadership. The SDWT synergizes the statements, "Break down barriers," "Rely on principles for direction," and "Use the frontline workers' skills and experience."

Quality by Design: Transition from Mechanistic to Organic Thinking

Ray Ford and his staff agreed that the frontline worker knows more about the job than anyone else. They also realized that Mountaintop's structure in 1991 inhibited performance and that it needed improvement. People wanted to do a good job, but the structure impeded them.

Harris' foray into Quality by Design showed the committee the power of teamwork. The staff saw that teams could

- Consistently solve problems more efficiently than individuals.
- Make higher-quality decisions than individuals.
- Handle more complex problems due to wider expertise, broad resources, and multiple viewpoints.
- Respond quickly and energetically.
- Divide tasks according to abilities.
- Increase individuals' commitment, motivation, and involvement.
- Promote individual skill development.
- Be more creative and innovative.

Ford and his staff next answered the question of what to change to. They decided to reorganize the frontline workforce into self-directed work teams. This arrangement expanded several job functions, cut redundancy, and flattened the organization. This restructuring coincided with a planned organizational downsizing, and was therefore less disruptive than it might otherwise have been. Engineering groups were also reorganized by process area. The reorganization got rid of the supervisory positions and redeployed former supervisors into technical positions. A single job classification now covered process technicians and maintenance technicians, who began the appropriate cross-training. These technicians joined production associates (operators) and indirect labor support (equipment repairers) by functional area (natural work groups) to create self-directed work teams. This reorganization flattened the organization and focused its attention on meaningful results (profit). The new structure appears in Figure 5.3.

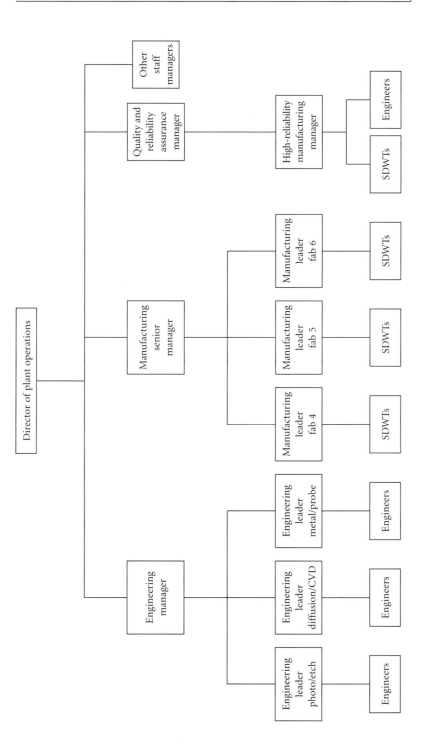

Figure 5.3. New Mountaintop organization: Self-directed work teams (simplified).

Open, Flexible Culture

The new culture also removed psychological barriers between leaders and followers. First, it got rid of preferred parking spaces for managers, which accords with Tom Peters' guidance. In 1996, when new construction limited the available parking spots, the management team reserved the better parking lot for the hourly workers (see Figure 5.4).

In 1993, a business-casual dress code replaced the shirt and tie. In addition, we established an open-door policy, encouraged calculated risk taking, and began to reward flexibility.

The Mechanism for Change

The staff next had to decide how to change. Chapter 4 says that there is no magic sequence an organization can use to make the necessary cultural change. They had to define the road map and steer the organization through it as necessary.

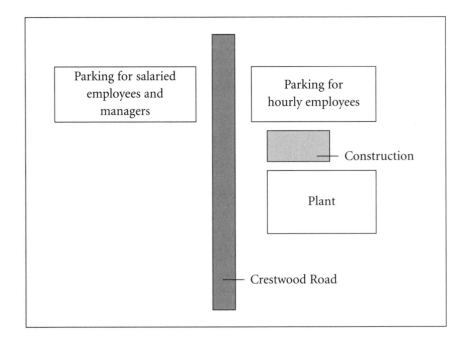

Figure 5.4. Parking arrangements at Mountaintop.

Peter R. Scholtes, in his excellent reference *The Team Handbook* (1988), defines a team as "a group of people pooling their skills, talents, and knowledge." The staff set out to define an SDWT and came up with this definition.

> *Self-directed work teams are natural work groups. Each SDWT is responsible for a business process that delivers a product or service to an internal or external customer. Each SDWT manages its business within defined boundary conditions. Its continued improvements for the success of the business are the measurement of its performance.*

Roles and Responsibilities

The staff had now redefined the organization and created SDWTs. The next task was to define roles within the new organization, beginning with its own. Staff members first established a plant steering committee, which consisted of a subset of the staff, union president Sally Hooper, and others. The steering committee's role was to

- Develop a structure that would align the organization's resources on continuous improvement in all areas and functions.
- Identify key business processes for improvement activities that would result in satisfied customers.
- Define Mountaintop plant goals that align with sector (business unit) goals.
- Champion Mountaintop's vision, strategy, and values.
- Define team boundary conditions.
- Review the progress of team improvement efforts and actively support constructive changes in the organization at all levels.
- Ensure adequate resources for continuous improvement.
- Champion improved communication throughout the organization.

The steering committee agreed to meet regularly. It also agreed that its composition should change over time to include people from other levels within the organization.

Role of the Team Member

The team member's role is to

- Use all of his or her available knowledge and skills to help define team goals.
- Be accountable for the team's success.
- Follow team agreements and hold other members accountable for them.
- Attend and actively participate in all team meetings and improvement efforts.
- Keep the team on track.
- Volunteer for assignments.

The plant reinforced the importance of teaming and the mutual commitment by labor and management. The following language became part of each job description during the next contract negotiation: "Team participation and the team approach to problem solving will be expected of all individuals in this occupation."

Role of the Team Leader

The team leader's role is to

- Be an active member of team.
- Monitor team progress.
- Interface with manufacturing leaders and other teams.
- Coordinate personnel/equipment scheduling according to team agreements.
- Ensure the availability of supplies by making requests to proper individuals.
- Use knowledge of the area to help prioritize equipment repairs.
- Prepare for and lead team meetings.
- Coach and encourage team members.

Team leaders do not receive extra pay. This was intentional. The steering committee felt that if additional pay were offered, two things might happen.

1. Some might take the job for the pay, not because they wanted to help change the culture.

2. Team members would perceive the leaders as pseudo-supervisors.

The plant defined the team leader role and provided adequate facilitation support to reduce the time spent on team leader duties. The first team leaders were salaried members, and the steering committee appointed them. This was necessary to provide leadership quickly: The salaried people all had received leadership training. After about a year, the teams could select any member as the leader. Within six months, 80 percent of the team leaders were hourly production associates.

Role of the Facilitator

The facilitator's role is to provide an immediate resource for team formation and development. The steering committee asked individuals to volunteer for the facilitators' team. These volunteers, from many different levels in the company, shared a common interest in making Mountaintop a better place to work. The volunteers received initial training in quality principles, problem-solving techniques, team dynamics, conflict resolution, and presentation skills. The steering committee then selected people from each shift to be the first Mountaintop facilitators. Members of this team would perform the role of a facilitator in addition to their current job responsibilities (production, maintenance, engineering, management). They would apportion their time accordingly, but never spend 100 percent of it facilitating. Like the team leader, they received no additional compensation for the job. Their role includes

- Helping the self-directed work teams mature and achieve goals through training, guidance, and coaching   ·

- Training and coaching in problem solving, group dynamics, team leadership, and meeting skills

- Helping teams interact effectively
- Offering support and encouragement
- Identifying and developing an awareness of specific behaviors that contribute to effective teaming and leadership
- Providing feedback based on systematic observation and listening

The Manufacturing Leader

The manufacturing leader became the first line of discipline for the teams. He or she is responsible for anything that does not fall under the role of the team, facilitator, or steering committee. Each manufacturing leader has between four and 12 teams (each team averages 10 members) reporting to him or her. The SDWT concept permitted the manufacturing leaders a *wider span of management* (as great as 100:1). Freedom from day-to-day supervision responsibilities lets them accomplish short-range objectives and participate in strategic planning. Their role consists of

- Performing all activities outside the teams' boundaries
- Interfacing daily with area teams and other manufacturing leaders and enforcing work rules for the area
- Reviewing team goals and needs for support
- Overseeing and directing staffing recommendations
- Giving public relations support and encouragement to teams

Source Matter Experts

The SDWTs consist primarily of manufacturing and support personnel. The steering committee had to create a role for people who are not SDWT members, but who work closely with them. These people usually provide technical expertise that is not available within the team. They assumed the role of source matter experts. Source matter experts may attend part or all of a team's meeting and receive time slots on the meeting agenda. Their responsibilities include

- Understanding and abiding by their role in the team meeting
- Understanding and responding to the team's need for information, guidance, or advice

- Informing the team leader of any changes related to their role and responsibilities with the team

Source matter experts come from areas like process or product engineering, facilities, and human resources.

Boundary Conditions

More words of wisdom: Self-directed does not mean anarchy, nor does it mean abdication.

To provide the teams with a working scope, the steering committee had to define some boundary conditions. They first agreed on a list of items that were outside the scope of the teams.

- Dismissing team members
- Disciplining team members—while self-discipline is the foundation of teaming, direct application of the four-step discipline process is outside the teams' boundaries
- Pay/bonus determination
- Formal peer appraisal
- Hiring—team members have participated in the interview process, but only a manager or leader can hire someone
- Budgeting—team members have input to the budget and responsibilities for controlling cost, but do not prepare the actual budget
- Direct management—that is, ordering another team member to do something
- Attendance records—a legal document requiring management confirmation
- Local/state/federal laws—team members cannot create any norms that violate these laws

The steering committee next defined some guidelines for teams' authority over the work. The committee first looked at technical decision authority (see Table 5.1). The committee also defined the team's administrative decision authority (see Table 5.2).

Table 5.1. Technical decision authority.

Issue	Team Authority
Dispositioning product	• Send on, rework, hold. • Scrap or release from hold with concurrence from engineering.
Equipment repair/ preventive maintenance	• Prioritize repairs. • Coordinate all repair/maintenance. • Contact and coordinate outside vendor. • Schedule preventive maintenance.
Equipment evaluation/ purchase	• Establish equipment needs. • Input on selection with engineering. • Input on performance acceptance specifications with engineering. • Evaluate new equipment with engineering.
Process improvement	• Collect data. • Utilize problem-solving techniques. • Sign off control chart with favorable response from corrective action. • Notify engineering of unresolved out-of-control situations. • Implement corrective actions.
Training	• Cross-training. • Who, what. • When—training methods to be coordinated with human resources and facilitators.
Supplies (e.g., chemicals, wands, gloves)	• Express need. • Evaluate alternatives.
Process setup	• Verify setup via statistical process control and operating instructions.
Hot lot expediting	• Red box priority unless overridden by manufacturing leaders. • Use of workstream dispatch where applicable.
Specification changes	• Recommend and review any changes with engineering and/or manufacturing leader. • Initiate documentation—specification action request, also known as engineering change notice.
Equipment utilization for engineering tests	• Evaluate impact on production schedule. • Coordinate engineering utilization with manufacturing leader approval
Stop processing	If the process is out of control, stop processing and contact the appropriate support group

Table 5.2. Administrative authority.

Issue	Team Authority
Product planning and scheduling	• Inter- (between) and intrateam resource allocation (people, supplies, equipment).
Goal setting	• The steering committee ensures that team goals aligned with plant and sector goals. • Set objectives to achieve team and department goals.
Team development	• Review performance goals as a team. • Periodically review individual team members for "areas for improvement." • Recommend corrective actions. • Seek assistance for performance improvement (from human resources and manufacturing leaders).
Physical environment	• Recommend and/or implement improvements on (notification to appropriate department necessary) —Housekeeping —Layout —Particle reduction —Safety • Interface directly with facilities support team when necessary

Steps in Forming a Team

The steering committee required a series of six steps to form an official team. With one minor change (to be discussed later), the plant uses this process today. The team can complete the steps in any order, but the following sequence has worked well.

1. Write and agree to a team charter.

2. Identify the team's customers and their needs.

3. Define team goals.

4. Write and agree to a code of conduct for the team.

5. Choose and agree on a team name.

6. Select a team leader.

The Team Charter

Write and agree to a team charter. A team charter states the reason for a team's existence. It can contain one or more phrases or sentences. A common form begins with the word *to* and continues with a phrase (or phrases). The words, "The charter of the _____ team is . . ." are implied to proceed the word *to*. Here is a famous charter that clearly defines an organization's reason for existence.

> *We, the people of the United States, in order to form a more perfect union, establish justice, insure domestic tranquility, provide for the common defense, promote the general welfare and secure the blessings of liberty to ourselves and our posterity, do ordain and establish this Constitution of the United States of America.*

A corporate charter will be less broad and sweeping, and a team charter will be narrower still. The basic idea, however, remains the same: The charter is the organization's reason for existence. The team should review its charter at least once every six months and whenever a new member joins the team. The team can rewrite it whenever it chooses.

Customer Identification

SDWTs are really customer-directed work teams. The concepts of customers and suppliers follow readily from the idea of a process. When people think of customers and suppliers, they usually think of external customers and suppliers. External customers buy the company's products, while external suppliers sell materials, information, or services to the company. External customers are the company's ultimate source of financial support, so we must keep them happy.

Customers and suppliers are, however, often internal. The people or organizations who precede our process are suppliers, and those who follow (or use our product or service) are customers. Employees pass work to other employees, who are internal customers for the work. Internal departments are often both customers and suppliers.

Each worker, therefore, is a customer of preceding workers. Each worker has customers: the people to whom the worker passes on his or her work. It is sometimes difficult to decide who is a customer (or supplier) and who is not. When there is no precise definition of customer, the problem usually lies in the definition of the process. For example, a project team discussing "increased safety" could never define a customer except in the most general terms ("everyone"). A project to "reduce hand and arm injuries during roll changes on machine number 6" focuses the definition. Team members would know that their customers are the people who use machine number 6.

To complete this phase of team formation, the team must identify its internal customers and identify the needs of each customer.

Goal Definition

Some authors subordinate goals to objectives, while others treat objectives as subordinate to goals—that is, goals support objectives, or objectives support goals. We will treat everything as a goal, with a qualifier. The sector bases annual goals on the strategic plan. These goals define the plant, or annual operating plan (AOP), goals. The AOP goals determine specific area goals. Each team must set goals every six months (each January and July) and submit them to the steering committee. Actions from team goals determine individual goals. Looking at it the other way, individual goals support the completion of team goals. Team goals support the achievement of area goals that in turn support plant goals, which in turn support sector goals (see Figure 5.5).

Team goals should serve the needs of the team's customers. A goal is a statement of results that the team wants to achieve. A goal should be elevating yet achievable. Goal statements should be clear enough to be understood by outsiders as well as team members. Lastly, a goal should be measurable. One goal format (the management-by-objectives goal format) is shown in the following examples.

> *Generic example: "To (action or accomplishment verb) (single key result) by (target date) at (cost)."*

> *Specific example: "To reduce scrap in the metals area by 50 percent by December 31, 19XX using existing resources."*

Another example is NASA's goal in 1961: "To put a man on the moon before the end of the decade."

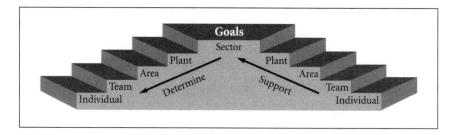

Figure 5.5. Goal alignment.

The Code of Conduct

Write and agree to a code of conduct for the team. A code of conduct describes how the team will operate and how the team members relate to each other. The code of conduct defines boundaries that the team imposes on itself. The team should review its code of conduct at least once every six months and whenever a new member joins the team. The team can rewrite it whenever it wants. For example,

- Be on time for meetings.
- Focus on the problem, not the person.
- Treat each other with respect.
- If you run out of work, help somebody else.

The Team Name

The team must choose and agree on a name for the team. This is a critical step in establishing an identity and becoming "us." A team can choose any name that is in good taste and that does not belong to another team.

The Team Leader

The Steering Committee defined three criteria for a team leader.

1. The team member must want to be a team leader.

2. The team member must be willing to attend training classes on how to be an effective team leader.

3. The team member must have satisfactory performance levels and must not be involved in any active disciplinary action.

Any team member, hourly or salaried, may (and should) take a turn serving as a team leader as long as he or she meets the criteria.

The team can select its leader in any manner that everyone agrees on. Some methods of selection include the following:

- Selection of a leader by lot.

- Candidates agree to take turns.

- Election by secret ballot or show of hands.

This exercise also served as the teams' introduction to consensus decision making. This decision, while not everyone's first choice, is one that all team members can support. Meanwhile, no one objected to it.

Mountaintop uses a term of six months for the team leader. Leader selection usually occurs in January and July to coincide with goal setting. The team may choose to adopt a shorter or longer term. If the team chooses (and the candidate still meets the criteria), the team may reelect a team leader to another term. In a strong team atmosphere, all team members should take a turn at being the team leader.

Organizational Focus

The new organizational structure supports the focus shown in Figure 5.6. The focus is on the AOP goals (the center of the circle). The self-directed teams make the goals' achievement possible, so they are the next highest level. Support organizations exist only to support the SDWTs, so they are third. The outer level depicts the steering committee. The teams on the right cross organizational boundaries to provide their support. The facilitators, for example, help break down interdepartmental barriers and make the organization porous. At the left are some TQM tools used in our environment.

Team Development

Larson and LaFasto (1989) chronicled their study of a "theoretically rich" sample of teams. Their goal was to answer the question, "What are the characteristics, features, or attributes of effective teams?" The study concluded that successful teams have the following characteristics.

- A clear and elevating goal

- A results-driven structure

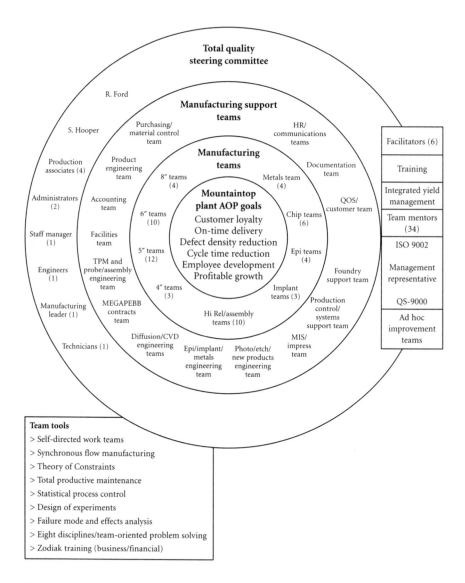

Figure 5.6. Mountaintop organizational focus.

- Competent team members
- Unified commitment
- A collaborative climate
- Standards of excellence
- External support and recognition
- Principled leadership

The following sections will discuss these characteristics and include examples from the SDWT environment at Harris Mountaintop.

A Clear and Elevating Goal

Teams define goals that are unambiguous, measurable, and constantly reinforced. During the first few years, the steering committee required each team to write goals every six months. Teams were to submit their goals to the steering committee, then report on their progress against the goals. They did this by providing a short written summary and an informal presentation. Upon completion of the goal, the team received a token, nonmonetary award. The teams initially chose simple goals, not directly aligned to plant goals. As time progressed, however, the goals became more challenging and in better alignment. Team awards progressed to plant awards. After a few years the steering committee (incorrectly) judged that it was no longer necessary to require the submission of goals. Many of the less mature teams gradually lost the habit of goal setting, and their development retrogressed. The required goal submission and alignment was reinstituted.

A Results-Driven Structure

Under this structure, teams organize, plan, schedule and document their progress. The steering committee set out to establish a team structure that would work in our environment. The teams were cross-functional, natural work groups, organized within their respective shifts. Decision authority and boundary conditions were clearly defined and communicated. Resources were made available as the need became apparent.

In 1995, the plant introduced shared responsibility for team management. The team leader served as a contact point and not an authority figure. Some team leaders, however, were beginning to carry a larger part of the load. At this point we introduced the "star concept" as an option for team organization.

Professional, civic, and fraternal organizations use the star concept, although they probably don't call it this. These organizations assign each critical task to a specific person: treasurer, secretary, newsletter editor, program director, and so on. This avoids burdening the president or chairperson (team leader) with all the work.

There is also an adage that "when everyone is responsible, no one is responsible." The story of Everybody, Anybody, Somebody, and Nobody illustrates this. The job was Everybody's responsibility, and Everybody thought Somebody would do it. Anybody could have done it, but Nobody did it. The key to shared responsibility is to assign every vital task to a specific person.

The steering committee defined the common responsibilities of a team as falling under five general areas: quality, safety, production, administration, and communication. If we assign a person (or persons) to assume the responsibility of a point (or points) of the star, they can make sure that goals are achieved and agreements are followed. I must emphasize that this does not mean that quality is only one person's job, or that safety is only one person's job. Everyone on the team is responsible for actions affecting each point. The "point person" is the focal point, as shown in Figure 5.7.

The team starts by listing the day-to-day activities for which it is responsible. Then, depending on the nature of the tasks, the team assigns them to points of the star. (A star can have any number of points.) The team considers the time necessary to accomplish the tasks and decides whether to assign one or two people. The team members agree on the assignment and rotation of duties.

The star relies on the "law of the situation" to achieve results. Juran and Gryna (1988, 22.60) say, "One *person* should not give orders to another *person*. Both should take their orders from the 'law of the situation.'" The task, or situation, becomes the authority figure. The situation's authority promotes compliance and mutual acceptance of responsibilities.

Competent Team Members

Team members need the right mixture of skills and a desire to contribute and work together. The team should not be too homogeneous, however,

Figure 5.7. Star organization for self-directed work teams.

or it will lose its creativity and versatility: "If everyone is thinking alike, no one is thinking." Mountaintop did not, unfortunately, have the luxury of selecting team members. Since the plant decided to change the entire organization at one time, there was no selection process. The cross-functional structure and natural diversity, however, ensured that the teams would not be homogeneous.

This presented a challenge to provide the team members with the skills necessary to interact as teams. The search for the "magic course" began. A few training courses were piloted, but none completely met the needs of the workforce. A thorough needs analysis showed that interpersonal skills were the most vital requirement. We found a good program from an outside training vendor, Zenger-Miller, and had them conduct a train-the-trainer session for our facilitators. The facilitators then delivered the applicable portions of the training to the teams, and then to the rest of the workforce.

We then addressed the next set of team needs: team leader training, meeting skills, problem solving, and teaming skills. In-house courses from Mountaintop and Melbourne filled these needs.

The maturity of the workforce, as well as the historically low turnover, also worked to our advantage. The close working relationships developed over the years helped reinforce the message that this was not "just another program."

Unified Commitment

Team members work to build team identity, unity, and commitment to their goals. Agreement on a team name is the first step in defining an identity. While team unity is necessary, we are careful to avoid suboptimization. We constantly remind teams that they are a part of the "big team." Winning the battle is useless if you lose the war. The chapters on synchronous flow manufacturing (chapter 9) and total productive maintenance (chapter 8) also discuss suboptimization. Performance measurements that focus on individual departments or activities in isolation are often responsible for driving the wrong behavior.

A Collaborative Climate

We want teams to focus on problem solving rather than finger pointing. The initial interpersonal skills training stressed the idea, "Focus on the problem/situation, *not* the person." We offer the following advice (source unknown).

> *Great minds discuss ideas.*
> *Mediocre minds discuss events.*
> *Small minds discuss people.*

Standards of Excellence

Under the Goldratt thinking process, teams must ask themselves what to change to. An answer often requires some form of benchmarking. We must ask the following questions.

- Where are we now?
- Where are our peers?
- Where are our competitors?
- Where do we want to be?

Standards of excellence are benchmarks for high performance. Self-directed work teams are *improvement* teams, not sustaining teams. There is always room for continuous improvement. Teams benchmark their own performance against other teams. The plant benchmarks our development against other organizations. Mountaintop employees have visited plants in the United States as well as in Japan. We've benchmarked other semiconductor companies, and manufacturing operations in other industries. Only by understanding what is considered world-class can we achieve that status.

External Support and Recognition

Management must support the goals, provide resources to accomplish them, and recognize their achievement. People will accomplish the goals only if management cares enough to ask about their status and make the necessary resources available.

There are many forms of recognition available to the employees at Harris Mountaintop. The site recognizes service anniversaries at regular intervals: 1, 5, 10, 15, 20, 25, 30, 35 and 40 years. Teams receive a small discretionary budget to have pizza parties a few times a year. When team leaders were the norm, we conducted half-day off-site team recognition events. These combined a brunch with some team-building activities. You'll read about the rewards for the zero scrap program in chapter 7. Here are two other forms of reward and recognition.

The *Harris Semiconductor Sector Awards Excellence!* program is a forum where any employee or team can reward any other employee or team for achievement in areas outside their normal job responsibilities. Categories include the following:

- Customer focus

- Continuous improvement

- Employee involvement

- Highest standards of business conduct

- Supplier partnerships

The steering committee reviews nominations and awards gift certificates for $25 (level 1) to the winners at the monthly communication meeting. The program also has provisions for level 2 and level 3 awards.

The "mother of all rewards" has to be TQM day. TQM day is an annual celebration of employee efforts. It is a thematic event, and the employees receive it very well. The first TQM day was in 1992, and was rather low-key compared to later celebrations. We set up a big tent on the front lawn outside the building and served hot dogs and soda during each shift for about two hours. We also awarded a few door prizes.

In 1993, our big focus was on total productive maintenance (see chapter 8). We held the same type of party (tent on the lawn, two hours per a shift), but established the theme "A Day at the Beach." Each employee received a beach towel with the TPM program logo. Bathing suits and beachwear were the attire of the day. Beach music played on the sound system while the teams competed in games related to the theme. Ray Ford and his staff took turns sitting in the dunking booth (see Figure 5.8). It was a hit.

In 1994, we had just received our ISO 9002 certification. We chose the theme "This is ISO Country!" Again the tent was set up on the lawn, only this time we had a full-blown country barbecue. Each employee received a

Figure 5.8. A day at the beach.

western-style neckerchief. Activities included line dancing and cow chip throwing (see Figure 5.9).

In 1995, a record year prompted an "Old-Fashioned 1890s-Style, Family Picnic" at a local country club. Employees and their families were invited. Activities included pony rides, horseshoes, a train ride, and more. We had face painting, a stilt walker, balloon animals, and an organ grinder and monkey. Entertainment included a Sousa-style brass band, a dixieland dance band, and even some karaoke. The highlight was the watermelon seed spitting contest (see Figure 5.10).

The next year, 1996, was the most profitable in the history of the plant. We had cut the ribbon on our $250 million expansion, Project Raptor. This led to the theme, "An Evening in Raptor Park." We rented a local armory and decorated it like Jurassic Park (see Figure 5.11). Each employee could bring a guest to the dinner dance. Casual attire was the rule, and several people came in costumes that supported the theme.

Variable Pay. Beginning with the labor contract effective December 1995, IUE Local 177 agreed to participate (as the salaried workers had previously) in a variable compensation program. A realistic profitability goal is established for the Semiconductor Sector at the beginning of the fiscal

Figure 5.9. Cow chip throwing.

Figure 5.10. Watermelon seed spitting contest at family picnic.

Figure 5.11. An evening in Raptor Park.

year. Employees receive a bonus, as a percentage of their salary or wage, for meeting or exceeding that goal. This percentage (of base salary) is paid out as a lump-sum bonus to every employee after the fiscal year ends. In fiscal 1996, this bonus was 4 percent (see Figure 5.12).

We must remember, however, the most effective recognition of all: a sincere "thank you" from one person to another.

Principled Leadership

The leader must provide enabling expectations for the team. Rogers and Ferktish (1996) put this very succinctly: "Empowerment doesn't release managers from accountability and responsibility. In fact, it places a greater demand on leadership than does the traditional organization."

History teaches us that when there is no strong leader, one will emerge. We have seen team leaders who perceived themselves as supervisors. We've

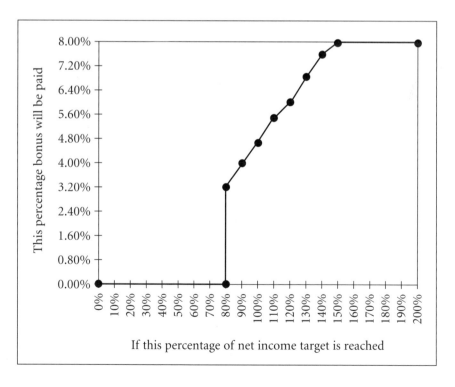

Figure 5.12. Variable compensation, 1996.

seen team leaders who were perceived as supervisors by the members. A vocal person would often assume the leadership role although someone else was the official leader. With the adoption of the star concept, many of these interteam struggles disappeared. The teams now look to their management and their union officers for leadership and guidance.

The teams studied by Larson and LaFasto (1989) were not self-directed work teams. They defined three basic team structures (see Table 5.3). Harris also recognizes three basic structures (see Table 5.4).

Communication

I know you believe you understand what you think I said; but I am not sure you realize that what you heard is not what I meant.

—Richard Milhous Nixon

Communication is the foundation of success. Miscommunication is a frequent cause of problems. Why is communication so difficult? Let's look at the elements of communication.

1. We begin with what a person thinks that he or she is saying (the thought they are trying to convey).

Table 5.3. Basic team structures (Larson and LaFasto 1989).

Broad Objective	Dominant Feature	Process Emphasis
Problem resolution	Trust	Focus on issues
Creative	Autonomy	Explore possibilities and alternatives
Tactical	Clarity	Directive Highly focused tasks Role clarity Well-defined operational standards Accuracy

Table 5.4. Basic team structures at Harris.

Type of Team	Role	Term
Project teams	Address a specific issue. Goals assigned by sponsor.	Disbanded upon completion of project.
Employee involvement teams	Perform similar or connected work within the organization; apply their knowledge and skill to effecting improvement in a defined area.	Long term in nature; new projects are defined as previous goals are met.
Self-directed work teams	Responsible for a business process, that delivers a product or service to a customer; manage their day-to-day operation; and effect continuous improvement.	A continuing function.

2. Next, we have what they actually said (exact words, placement, emphasis).

3. Finally, we have what the listener thinks they said.

Why aren't these three items the same? You may have played the parlor game where one person whispers a short story to the next, who quickly whispers it to the next, and so on through a group of people. The last person repeats the story aloud. It usually bears little resemblance to the original. This is the communication paradox at work. Everything we say and hear is filtered by our own ideas, opinions, beliefs, paradigms, and values (see Figure 5.13).

Improving Communication

At Harris Mountaintop, we are always trying to improve our communication. Employees can approach managers of all levels. Each team has its own mail slot and an e-mail account. There are traditional bulletin boards and scrolling electronic displays in key areas. The steering committee publishes a newsletter that alerts employees to major changes that affect the organization.

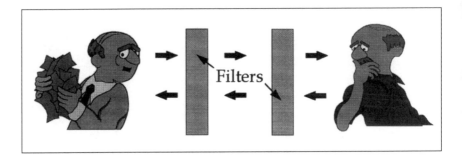

Figure 5.13. The communication paradox: What you say may not be what is heard.

The most significant event is the monthly communication meeting. This is a monthly 30- to 45-minute all-hands meeting for each shift. The plant manager or a staff manager shares the state of the business and summarizes performance against the AOP goals. The speaker presents expectations, discusses key ideas, and explains coming events in detail. A question and answer period follows. The president of the sector holds a semiannual communication meeting at the plant to share progress on a sector level. A typical communication meeting is pictured in Figure 5.14.

Figure 5.14. All-employee communication meeting.

Fab 8: Doing Whatever It Takes

When challenges arose during the initial years of teaming, people made many suggestions to overcome them. We might have better teams if

- Teams could choose their members.
- People received team training before forming teams.
- People could form teams and get to know each other before working together.
- Facilitation support was adequate.
- The system supported team development.

The chance to try these ideas came with our most recent expansion.

In 1996 Mountaintop began staffing its newest fabrication facility. Fab 8, or Project Raptor, is the world's first eight-inch (200 mm) discrete power pipeline. (Fab numbers refer to the wafer size. Fab 8 processes eight-inch wafers. See http://www.mtp.semi.harris.com/raptor.html for more information.) It represents Harris Corporation's largest single investment: $250 million. Harris structured Fab 8 for success throughout all aspects of its planning and construction.

The manufacturing technology in Fab 8 required new hourly occupations: fabrication specialist and fabrication support specialist. Fab 8 team members were then carefully chosen through an internal interview process. Criteria included attitude, flexibility, and performance record. The Fab 8 team took a cue from Nike's motto, "Just Do It!" to develop the motto, "Whatever It Takes!"

Training

After the plant selected the team members, it had to train them in the necessary team skills. A two-month training period would precede the factory startup. The training included team training as well as technical and equipment-specific training. The composition of the four Fab 8 teams is unique to Mountaintop. Each team is structured around the entire pipeline concept, and each has its own support personnel. This means that each team includes fabrication specialists from each process step, fabrication support specialists, technicians, a mechanic, an electrician, and a utility associate. Each team works on a compressed schedule and can run the entire process from start to finish. Harris Mountaintop's other self-directed work teams are in single, focused work areas. The support teams are separate entities. The unique characteristics of the Fab 8 teams required careful design of their training schedule.

Most of the people chosen for Fab 8 came from the existing workforce, and therefore had learned teaming basics. The challenge was to select and develop the necessary courses and deliver them during the two-month training period. The plant hired additional facilitators/trainers to accomplish this. They delivered the following courses, in addition to equipment-specific training.

- Advanced teaming workshop: 4 hours

- Interpersonal skills: 4 hours

- Effective meetings: 4 hours

- Problem solving techniques: 8 hours

- Theory of Constraints (production application): 16 hours

- Statistical process control: 6 hours

- Windows operating system and cc:Mail (communication): 8 hours

Team Formation

Actual team formation took place in three two-hour sessions. During the first two sessions, each team, aided by a facilitator, completed the formation steps summarized earlier. The only step that changed was "selecting a team leader." This step has become "choosing a structure." The teams have the option of choosing a team leader structure or the star concept. All four Fab 8 teams chose the star concept.

The teams had the opportunity to use the skills from their training: how to have effective meetings, customer/supplier relationships, brainstorming, and consensus decision making. The teams developed charters, agreed on codes of conduct, identified their customers and suppliers, and structured their goals.

During the third formation meeting, the teams discussed how to best use the 45-minute crossover period in the Fab 8 schedule (another Mountaintop first). The crossover allows the shift that is leaving to meet with the shift that is coming to work. The two crews involved in the crossover met to discuss this as a group. Therefore, crews A and C met as a group, and crews B and D met as a group in two different sessions.

Both sessions followed the same format: The two crews broke up into four brainstorming subgroups, each with members from both crews. The four subgroups brainstormed crossover agenda items, crossover location, and attendees. Each subgroup presented its ideas, and the two entire crews discussed the ideas to arrive at a consensus. There was a followup meeting of all four teams to set up a uniform crossover format.

The plant selected Fab 8 team members carefully, and the system design supported teaming. The plant hired additional facilitators, and training and team formation took place before the new factory opened. Mistakes are acceptable if they promote learning and improvement. Fab 8's success will testify to how well we've learned from our mistakes and how far we've come so quickly.

Frequently Asked Questions

You can always spot Internet users because they add a FAQ section in everything they write! Many groups have visited Mountaintop during the past few years to benchmark our self-directed work teams. These groups invariably ask questions like the following, so I will provide the answers here. Happy teaming!

Q: Do you believe that self-directed work teams were solely responsible for a $50 million turnaround in five years?

A: Three factors were responsible for the turnaround, but teams were the foundation. The market for our products improved during the past five years. Luck, however, is the meeting of preparation and opportunity. The quality and customer focus of the teams was the preparation. Quality and productivity programs do not function in isolation; they support and augment each other. The teams enabled manufacturing improvements such as total productive maintenance, synchronous flow manufacturing, and integrated yield management. These improvements turned market opportunity into profit.

Q: Doesn't the time taken for meetings lower the productivity of the workers?

A: No. Usually the improvements developed by the team far outweigh the time spent to develop them.

Q: What about the cost of training these people? What is the payback?

A: We look at training as an investment, not a cost. Technical training is often more applicable to improvement activities than soft training. However, without investing in organizational development, the technical training will be less effective. You sometimes do things because they are the right thing to do. The results are visible in the increased maturity of the teams and in the bottom-line financial results.

Q: How much training do you do?

A: The correct answer is "as much as we need." You are probably looking for numbers, however. We want an average of 24 or more hours of formal (classroom) training per person per year. This is only 20 percent to 40 percent of all training. On the job training is 60 to 80 percent.

Q: If there are no supervisors, who handles discipline?

A: The nature of the teams is that of self-discipline. With empowerment comes responsibility. This is not to say that some people don't try to see what they can get away with. Those people exist. Those people also existed when we had supervisors. The difference is that the supervisors spent 80 percent of their time dealing with those 20 percent of the problems. Now everyone can use that time more productively.

Q: So is there no formal discipline?

A: There is. We have a standard four-stage discipline system: verbal warning, written warning, suspension, then termination. Discipline is administered by a manufacturing leader or manager.

Q: Would you recommend teaming for my organization?

A: It worked for us, and it will probably work for you. It will take patience, commitment, and effort. Don't copy the way we've done it, since our approach is unique to our organization. Study what we've done and adapt it to your situation. Don't start, however, unless your management team is absolutely committed. There is an old analogy that compares involvement and commitment: bacon and eggs. The chicken was involved, but the pig was committed.

Management Commitment: The Vital Ingredient

Stephen Covey's "Law of the Farm" (1991) describes the nature of this commitment. Many organizations want to buy an instant quality program that will give them results in a month or two. They expect, like Jack in the fairy tale, to plant a bean one day and have a giant beanstalk the next.

Many management teams launch quality improvement programs without sincere commitment to their success. The programs' initial fanfare raises employees' hopes, enthusiasm, and expectation. When management abandons the programs or lets them die through lack of support, the employees feel disillusionment and demoralization. They will be less likely to believe in the next program that comes along. The practice of generating enthusiasm for an activity, and then not following through, is a psychological weapon. In American football, the defense may call a time-out before the opposing kicker tries a field goal. The kicker is mentally "up" for the kick, then he has to wait.

Many people use this weapon on their own organizations! Armand Feigenbaum (1991, 193) says that lukewarm support for a quality program is the kiss of death. In summary, *don't blow the trumpets unless you mean it. Unless you plan to follow through with the program, whether it be teaming or something else, don't start.*

Covey (1991) compares business management, and life, to a farm. We must plow and sow in the spring, and tend the crops all summer, before we can harvest in the fall. Success requires effort, persistence, and patience.

CHAPTER 6

Customer Contact Teams: Improving Communications and Quality

Allen Sands

Harris Semiconductor's customer contact teams (CCTs) promote customer-supplier communications at the shop floor level. Most CCT members are manufacturing workers, and they talk directly to the customer's frontline workers. CCTs solve problems and deliver customer-specific quality improvements. They rely on three characteristics.

1. They use the frontline manufacturing workers' knowledge, skill, and experience.

2. They open short, direct communications between the people who make a product and the people who use it.

3. They improve sensitivity toward customer concerns within the organization.

CCTs also promote morale and intrinsic motivation. The CCT provides large doses of the elements vital to intrinsic motivation: skill variety, task identification, task significance, autonomy, and feedback. The CCT asks the supplier's and customer's workers to think, identify problems and opportunities, use judgment and initiative, and offer solutions.

The CCT's Role in a Dynamic Competitive Environment

Crisis is Chinese for "danger and opportunity." This aptly describes today's business environment. "Violent and accelerating change, now commonplace, will become the grist for the opportunistic winner's mill. The losers will view such confusion as a 'problem' to be 'dealt with'" (Peters 1987, 14). Two characteristics of CCTs make them very effective tools for prospering in this environment. They use the frontline worker's skills and experience, and they shorten the customer-supplier communication pathway.

The Role of the Frontline Worker

The resources available to a business corporation include its people. Using them effectively can provide a decisive advantage over a competitor with equal or even superior physical resources. Armand Feigenbaum (1991, 207) writes, "the most underutilized resource of many companies is the knowledge and skill of employees." W. Edwards Deming said, "the greatest waste in America is failure to use the abilities of people" (Covey 1991, 264).

Tom Peters (1989) says that the frontline worker knows more about a job than anyone else. He or she handles the production equipment, raw materials, and subassemblies every day. While the manufacturing worker may lack a college degree, engineers and managers do not have the worker's extensive, detailed, hands-on experience. From a management or engineering perspective, incoming materials and parts must meet certain specifications. Frontline experience may show, however, that the specifications do not guarantee that materials will work under manufacturing conditions. Therefore, frontline worker involvement can often produce additional solutions or improvements.

Porous Customer-Supplier Relationships

W. Edwards Deming said to break down barriers between departments. The CCT extends this guidance to customer-supplier barriers. It makes the supplier, to use Tom Peters' term, *porous* to customers. Peters (1988) says that porosity makes organizations flexible and responsive to diverse customer needs.

This principle applies to all forms of information transfer. Figure 6.1 shows the traditional customer-supplier communication chain. Suppose the customer's workers are having trouble with a supplier's parts or materials. They complain to their manufacturing engineers or managers. The latter pass the complaint on to purchasing. The purchasing department contacts the supplier's sales department. The sales people contact their own manufacturing engineers or managers. The latter may or may not discuss the problem with the frontline workers. They may not understand why there is a problem if the shipment met specifications. There are two deficiencies in this communication process: (1) it is slow because information must pass through several departments; and (2) the process loses information because the people who use the product do not talk to those who make it.

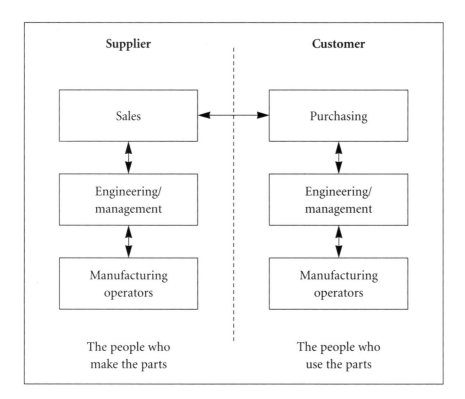

Figure 6.1. Traditional customer-supplier communications.

Peters (1988, 14) specifically advocates having frontline workers interact with customers and suppliers: "the 'average' person . . . will routinely be out and about—that is, first-line people communicating directly with suppliers, customers, etc. Who is the person who best knows what's wrong with defective suppliers? Obviously, the frontline person who lives with the defective item eight hours a day." Peters also says that frontline workers can work with suppliers to improve quality or productivity (p. 18).

Getting Ideas from Customers

Harris Corporation's first customer contact project started in 1898. The company was not then in the electronics business; it sold printing equipment. Charles and Alfred Harris sold an E-1 Harris Automatic envelope feeder to the Enterprise Printing Company in Columbus, Ohio. The machine's function was to feed envelopes into printing presses.

Alfred Harris was installing the E-1 on the customer's shop floor when a printing error happened at a nearby workstation. A worker who was hand feeding sheets into the printing press missed a sheet. The pressman showed visible dismay. Harris did not understand why, so the pressman ran another sheet through the press to show what had happened. Because of the missing sheet, the press had printed the image on the impression cylinder. The cylinder then placed a reverse image on the next sheet that went through the press.

Harris and the pressman looked at the sheet together. They noticed that the lithographed image on the back of the sheet was better than the one on the front. The pressman exclaimed, "If only we could turn out litho like that on the *front!*" This observation by the customer's frontline worker inspired Alfred and Charles Harris to develop an offset lithographic press. This went to market in 1906, and the Enterprise Printing Company was one of the buyers (The inspiration for the offset press 1995).

The Customer Contact Team

Harris Semiconductor's plant in Mountaintop started its first contact CCT about three years ago. Workers at Customer A's plant were complaining about Harris parts. The parts met the customer's specifications, but caused stoppages in the production equipment. The communication pathways in the traditional customer-supplier relationship did not provide the information that Harris needed to deal with the problem.

Team Composition

Harris assembled a team of hourly production workers, engineers, and managers. The production workers were volunteers from teams whose activities could affect the customer's problem. A typical work team at Harris has between five and 10 members. There were more volunteers than places on the CCT, so lots were drawn for the CCT positions.

Team Preparation

The production workers received some basic training from the site quality and customer satisfaction manager on how to work with customers. The teams also learned a basic version of Ford Motor Company's TOPS-8D (team-oriented problem solving, eight disciplines) technique. This systematic problem-solving method features these key instructions.

1. Use a team approach.

2. Develop a working definition for the problem. This includes comparing what should have happened with what actually happened. Also, where in the process does defect detection happen? Which step (and which workstation) probably generated the defect?

   —Juran and Gryna (1988, 22.35) point out that an ambiguous or imprecise problem statement often impedes the problem-solving activity. This element of TOPS-8D helps ensure a clear problem definition.

3. Implement a containment action (fix the problem's symptom) and make sure that it is effective. Containment includes dispositioning the nonconforming product and preventing further manufacture of bad units.

4. Identify the problem's root cause.

5. Select and verify a permanent correction for the root cause (fix the problem).

6. Carry out the permanent correction and monitor the action's effectiveness.

7. Prevent the problem from recurring.

8. Recognize the team's achievement.

TOPS-8D contains all the elements of the plan-do-check-act (PDCA) quality improvement cycle. Juran and Gryna (1988, 10.25–10.26) describe PDCA, which includes using a control cycle to make sure that improvements become permanent. Step 7 of TOPS-8D addresses the latter element.

Team Visit to Customer A

The team visited the customer's plant and met with the manufacturing workers there. This face-to-face contact with the product's users opened a short and direct communication pathway at the shop floor level (see Figure 6.2).

This innovative approach quickly revealed the problem's root cause. The surface color and reflectivity of the Harris semiconductor chips were causing trouble for the customer's machine vision systems. The specifications did not cover reflectivity or color. Harris' team solved this problem by making the color of the chips lighter and more consistent.

This experience illustrates the concept *fitness for use:* "'fitness for use,' 'inferior,' 'competitive,' and 'superior' all relate to the situation *as seen by*

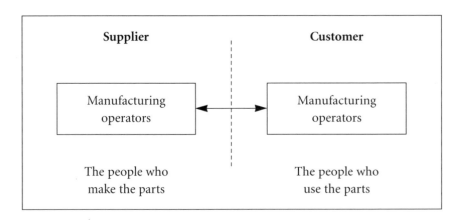

Figure 6.2. Customer contact teams.

the user" (Juran and Gryna 1988, 3.9). Fitness for use means the ability to meet all aspects of a customer's needs. Customers express needs through specifications, but specifications do not always reflect every need. Therefore, meeting specifications does not ensure fitness for use or customer satisfaction. Juran and Gryna emphasize the difference between conformance to specifications and fitness for use (3.6–3.7).

What started as a response to a customer complaint continued as a joint effort to further improve the customer's productivity. Semiconductor chips (or die) come on circular wafers. An electrical tester at the end of the manufacturing process marks the nonconforming ones with ink. A saw separates the die, and a picker (see Figure 6.3) lifts them from an adhesive film. Automatic picking equipment will select the good chips and leave those with ink dots behind.

Harris normally places the good die on tape reels for shipment (see Figures 6.4 and 6.5). However, Customer A buys die on wafers instead of reels. The customer's die bonders transfer the die from the wafers to copper heat sinks. Skipping the bad die reduces this equipment's speed. The CCT devised a process to take good die and put them on Mylar tape in the

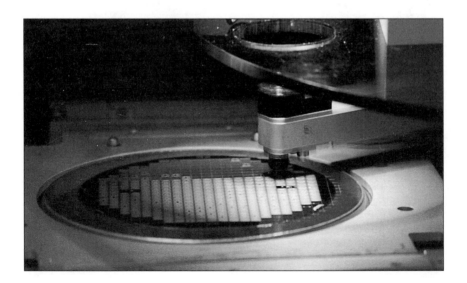

Figure 6.3. Automatic picking equipment: The wafer is on Mylar film, and a diamond-edged saw has separated the die.

Figure 6.4. Standard process: Placement of die on tape reel.

Figure 6.5. Die in tape reel: Standard process.

shape of a wafer (see Figures 6.6, 6.7, 6.8, and 6.9). Since all the chips are good, the customer's equipment can run at its top speed.

The shop-level interaction also identified an opportunity to improve the solderability of Harris' chips. Finally, Harris operators taught the customer's workers how to avoid damaging the chips with contamination or electrostatic discharge. This approach's decisive success led Harris to extend the CCT to other customers.

Visits to Other Customers

A CCT visit to Customer B led Harris to redesign packages to prevent product spillage during shipment and handling (Reaching out to customer 1994).

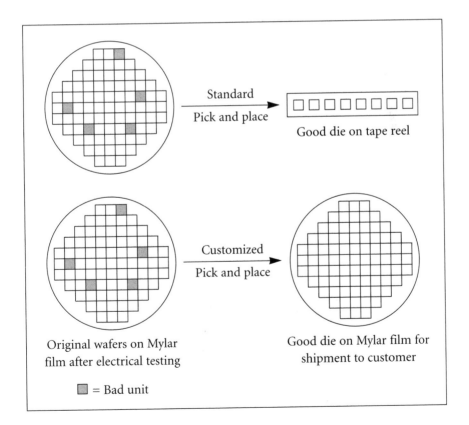

Figure 6.6. CCT's innovation to improve productivity.

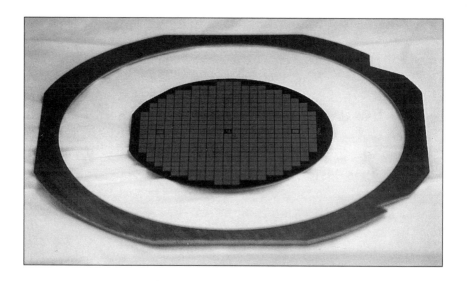

Figure 6.7. Wafer on Mylar film, ready for dicing and picking.

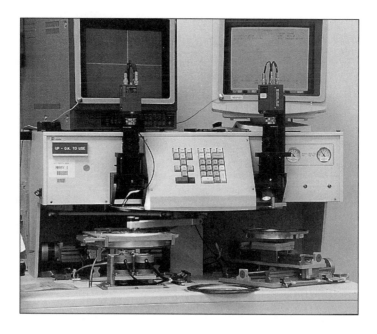

Figure 6.8. Customized process: Die from diced wafer to Mylar tape. The arm takes die from the diced wafer (left) and places them on Mylar tape (right).

Figure 6.9. Customized process: Placement of good die on Mylar tape (closeup of right-hand side of Figure 6.8).

Production workers at Customer C said that more die per tape reel would improve their productivity. Harris increased the reel size by 67 percent. This change increased the time between equipment setups (reel changes) (Mountaintop team visits 1994). Harris was unaware of this consideration before the CCT visit, but it affected the customer's purchasing decisions. The customer's shop floor personnel favored a competitor who offered more die per reel. The CCT again supported the basic rule for winning and holding market share: *Know what the customer wants and deliver it.*

Experience to Date

Harris' customer contact teams have visited six customers. One visit was enough to achieve results for five customers. There have been several visits between Mountaintop and Customer A. We have seen that the customers, including their frontline workers, have been very happy with these visits.

Why Does It Work?

Workers feel more comfortable discussing issues with like-minded individuals who understand the manufacturing environment. This comfort level promotes open, trusting communications.

Open discussions between Harris' workers and the customer's workers created a peer environment. This environment allowed day-to-day problems to come out naturally, because frontline workers are intimately familiar with them. This on-the-job, team environment let Harris' workers expose floor-level issues that were not apparent at the technical level.

Also, Harris' workers wanted to understand first-hand the customer complaints. By doing this, they felt more in control of the solutions. A feeling of control is a feeling of ownership.

Obstacles Encountered

Most new programs and methods run into resistance, and the CCT was no exception. Customers took the traditional position that it was the supplier's job to correct problems or improve the product. They soon recognized, however, that customer-supplier cooperation is mutually helpful.

Some of Customer A's technical personnel believed in the extremist form of the Taylor management model (see Figure 6.10): Engineers and managers do the thinking and planning, and hourly workers do what the thinkers and planners tell them. They didn't believe that hourly production workers could solve a technical problem. They thought the CCT activity would be a waste of time. The CCT's success quickly convinced them that daily experience with a job provides valuable practical knowledge.

Open and frank communications on the factory floor between Harris' workers and the customer's workers generated trust and excitement. This environment helped reveal issues that an engineering-level discussion might not have detected.

The CCT also supported both companies' focus on the teaming aspect of total quality management. After the visit, Harris' management

All employees: Leave your brain here
before starting your shift.

1   2   3   4   5

Brains by Corel Clip Art

Figure 6.10. Taylorism taken to an extreme.

encouraged CCT members to present their findings to fellow work team members, other teams, and the plant steering committee. These discussions went beyond problem resolution, and addressed team skill benchmarking for both Harris and the customer. These results convinced the customer that the CCT was a good idea. This led to return visits and exchanges, where the customer's manufacturing team members visited Harris.

Task significance is an intrinsic motivator. It means that workers know how their activities affect their customers. CCT activities improved the team members' understanding of the customer and its application of Harris' products. This understanding dramatically improved their sensitivity to the customer's needs, which made their work more meaningful and satisfying.

There were concerns about the production workers taking time from their regular assignments to work on CCTs. The plant's management team recognizes that customer satisfaction *is* part of the production workers' job. Assigning the workers to CCTs instead of routine production was an investment in customer satisfaction.

Table 6.1. Motivational elements of CCTs.

Motivation Element	Description	Role in CCT
Skill variety	The job exercises many of the workers' skills and talents.	The CCT asks the supplier's and customer's workers to identify problems and opportunities and to offer solutions. This requires them to apply a wide range of skills and experience.
Task identity	The job has a visible and identifiable outcome. This is the antithesis of the image of the workpiece as a meaningless "widget."	A semiconductor wafer or chip fits the stereotype of a widget. The CCT, however, allows frontline workers to see how the external customer uses the final product.
Task significance	The workers can see how the job affects internal and external customers.	The workers understand the product's importance to the customer. If defective product causes excessive downtime, the customer's employees may lose work and pay. The product's quality affects the customer's costs and profits.
Autonomy	The workers can exercise judgment and initiative in planning and carrying out the work. Autonomy is the opposite of the "leave your brain at the factory gate" legacy of Taylorism.	CCT activities *require* team members to think and to use judgment and initiative.
Feedback	The job provides feedback on the results of the workers' performance.	CCT members gain satisfaction from seeing their employer use their ideas. They also receive feedback on how their contributions affect the customer's satisfaction.

Motivational and Morale Benefits

Extrinsic motivators like pay and benefits cannot evoke the consistent attention to detail that is the human factor in quality. Only intrinsic motivation, which makes the task its own reward, can do this. The CCT provides large doses of the five elements vital to intrinsic motivation: skill variety, task identity, task significance, feedback, and autonomy (Schermerhorn, Hunt, and Osborn 1985, 211–212). Table 6.1 shows the motivational elements of CCTs.

Customer A's production workers said they enjoyed working with Harris' CCT. One said that the manufacturing operators who made the product were in the best position to understand their needs. Another gave a suggestion to the CCT, which took it back to Harris for implementation. Customer A sent a team to visit Harris' Mountaintop facility, and invited the Harris team for a return visit.

Conclusion

Harris Semiconductor's customer contact teams have produced major quality and productivity improvements. The CCT provides a short communication pathway and involves the frontline production workers. The program has been so effective that Harris is actively expanding it. CCTs have visited six customers, and the company plans to extend the program. The CCT provides morale and motivation benefits to both the customer and supplier. The CCT has proven itself a progressive tool for competitiveness in today's dynamic marketplace.

CHAPTER 7

Zero Scrap Actions

Allen Sands

Harris Mountaintop's zero scrap activity relies on the self-directed work teams (SDWTs) to reduce scrap, rework, and defects. Teams don't need extensive technical help to reduce scrap, although this is a common paradigm (see Figure 7.1). Mountaintop's experience is that teams know about and control many factors. Most teams know where product losses happen in their areas and why. With a little encouragement and a small investment, team-initiated projects can bring major scrap reductions.

Introduction

Two types of fallout occur during wafer processing. First, human error or equipment problems may break the wafers. Second, technical problems may cause the wafers to become scrap.

Breakage is clearly observable by the manufacturing team members. Equipment may function poorly because of poor preventive maintenance or design problems.

The process itself may be conducive to human error. Several years ago, the Harris facility in Mountaintop was losing wafers to "operator error." The operators handled wafers with tweezers and often dropped them. Tweezers had for many years been accepted for handling wafers, but now there was a breakage problem.

The semiconductor's first silicon wafers were an inch (25 mm) in diameter and very light. Even a three-inch (75 mm) wafer is light. Wafers

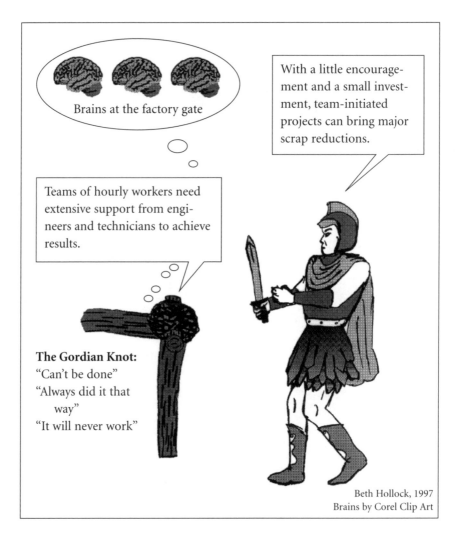

Figure 7.1. The frontline worker's role in scrap reduction.

now range from four to six inches (100 to 150 mm), and Mountaintop has built a plant to process eight-inch (200 mm) wafers. As wafers become wider, they must become thicker to support their own weight. The new wafers are much heavier than the older ones and are difficult to hold with a tweezer. If the operator does not squeeze hard enough, the wafer will fall on the floor and break. If the operator squeezes too hard, the pressure may crack the wafer. The problem was really inappropriate tooling, not "operator errors." When Harris bought vacuum pencils for wafer handling, the operator errors went away. A self-directed work team initiated this change.

In a wafer fabrication line (fab), several common technical problems can occur. These include equipment malfunctions, and failure of visual or electrical criteria at inspection and test operations. These inspections and tests often involve statistical process control (SPC) checks. The bad product, however, not only violates statistical limits, but exceeds the specification. Despite the problems' seemingly technical nature, team members have a good idea of where repetitive fallout or scrap occurs.

The team members are in the best position to identify sources of wafer fallout. This book often repeats a key lesson: *The frontline worker who does the job eight or more hours a day knows more about it than anyone else.* General Patton said that frontline soldiers know more about the war than anyone else. Tom Peters drives home this lesson repeatedly in his books and lectures. Dr. Stephen Covey says that workers who know the company's principles can take extensive responsibility for quality and customer satisfaction. Harris' Mountaintop plant has applied this lesson to its everyday operations.

The team members are in the best position to know which equipment breaks wafers and which processes don't work well. To be useful, however, breakage, scrap, and rework data must be available to the team. The next section shows how Harris Mountaintop put these principles to work to reduce scrap and improve quality.

Mountaintop's Zero Scrap Program
We will look at how the teams reduced scrap in their work areas. The following methods generated 100 completed scrap reduction actions and projects.

Effective Information
Management must deliver effective information to the teams. The team members know where they reject wafers, but they need information so that they can quantify problems. Mountaintop uses Consilium's Workstream production control system to track wafer losses.

Measurement/Pareto Chart

The teams can use the loss data to develop a Pareto chart. The Pareto Principle says that 80 percent of the trouble comes from 20 percent of the sources. The Pareto chart is a histogram that ranks the sources from largest to smallest (see Figure 7.2). This helps the team focus resources to achieve the largest improvements. The data also let the teams quantify their achievements in the language of management: dollars.

Mountaintop's operators receive training in SPC. The course equips them to use Pareto charts and apply the Pareto Principle.

Feedback, or letting people see the results of their efforts, is a vital intrinsic motivator in any job. Measurement systems that help the teams track and observe themselves encourage the teams to use the improvement techniques. Effective scrap tracking closely coupled to the team's natural work flow is very useful if it is readily available.

Quick Fixes

As the teams start this process, there will be some obvious "quick fixes" for some problems. The teams know these from daily experience and intuition. If management gives them the freedom to do so, they will overcome these problems immediately. Here are some examples.

1. Photolithography defines the microscopic features on silicon wafers. The first step is to coat the wafer with a photosensitive polymer, or photoresist. An exposure tool, or aligner, exposes this coating to a pattern on a lithographic mask. The exposure tool is like a camera, albeit a very sophisticated one. The features of successive layers must line up properly, and the aligner should line them up automatically. The teams discovered that specific aligners did the best job on certain products. The factory has assigned designated aligners to these products.

2. An improvement in the staging area reduced misprocessing incidents.

3. Improvements in wafer transfer methods and equipment reduced breakage.

A string of successes, whether small or large, builds a team's morale. There must be an effective feedback mechanism, since the teams need to see the fruits of their activities. The teams' actions produce improvements they can celebrate, which builds their confidence.

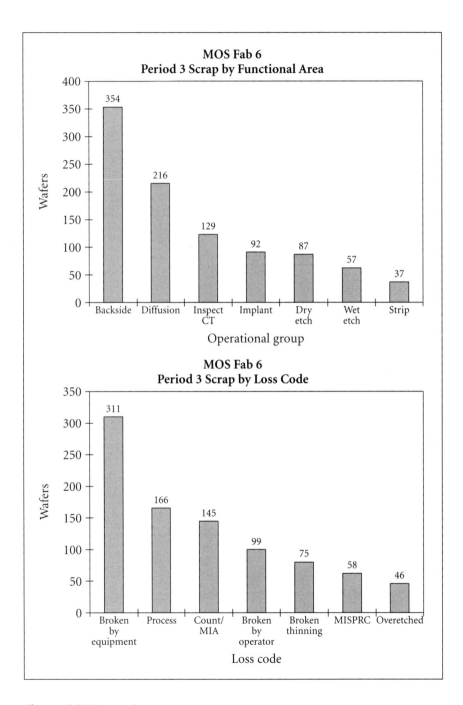

Figure 7.2. Pareto charts.

Communication and Visibility

There must be a visible way to track the actions. At Mountaintop, we graphically tracked wafers with zero yield at probe. What is wafer yield? A semiconductor factory manufactures transistors and diodes on silicon wafers. Each electrical device is a tiny rectangular piece of silicon. These pieces of silicon are die, pellets, or chips. At the end of the process, a saw cuts the wafers into individual die. A wafer can supply several hundred, or even thousands, of die. The yield is the portion of the die that pass the final electrical test (see Figure 7.3).

Zero yield means the wafer is dead on arrival at final test: There are no good die on the wafer, and the factory wasted the resources that went into making it. Chapter 9, on synchronous flow manufacturing, will show that the real cost far exceeds the wafer's book value. The real cost is the opportunity cost, or marginal revenue from selling another wafer. When the factory is working at capacity, a dead wafer represents an irreplaceable loss.

Therefore, whatever the teams did to reduce the incidence of zero yield wafers was very profitable. Mountaintop posted the monthly zero yield results in the cafeteria, and management reviewed them at monthly employee communication meetings. We also set up a "Road to Zero Scrap" and paved it with the teams' scrap reduction activities. Pride developed as the teams presented their actions at monthly all-employee

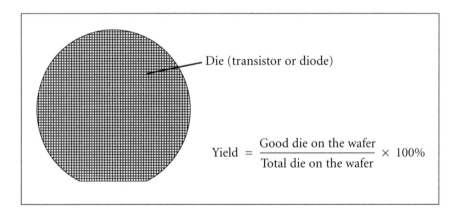

Figure 7.3. Silicon wafer, die, and yield.

meetings. These teams received reinforcing feedback, and their successes encouraged other teams.

Action submission sheets were readily available to all teams (see Figure 7.4). When a team completed an action, it filled out a form. The team described the action and estimated the monetary savings. Again, the availability of scrap information and financial data helped the teams quantify their results. Two members of the site steering committee reviewed the submissions. Approval of the actions was generally prompt, and the reviewers tried to err toward approving any reasonable action.

Reward and Recognition

The Mountaintop steering committee arranged three reward and celebration incentives to promote the scrap reduction program. Reward and recognition systems can vary by individual industries and circumstances, but they must capture the teams' interest and show management's appreciation.

1. The first goal was to achieve 100 actions for scrap reduction. When the plant achieved this, management arranged a party to celebrate the milestone. The event recognized the entire plant's contribution to scrap reduction.

2. If a team's actions produced exceptional savings, a member of the steering committee could nominate the team for a special reward. This was usually a recognition dinner for the team members. These separate celebrations were for actions clearly above the norm, with demonstrable savings.

3. Every successful project earned its team a ticket for a drawing. The teams with the most projects had the best chances of winning. The three winning teams sent two of their members to Harris facilities around the world. One trip was to Kuala Lumpur, Malaysia, the site of Harris' largest assembly facility. Another was to Dundalk, Ireland. The third team's representatives went to domestic sites in Findlay, Ohio, and Melbourne, Florida. The representatives visited the sites and discussed Mountaintop's scrap reduction program. They also explained Mountaintop's team structure and did some benchmarking teaming with their hosts. There was an opportunity for some sightseeing around the host plant. The prospect of visiting our sister plants had widespread appeal.

Zero Scrap Program

The Road to Zero Scrap

Team name	
Shift	
Originator	
Cooperating individuals	
Description of scrap reduction	
Date implemented	
Approximate savings: Wafers, die, money (if available)	

Approval Date: Signed:		Number

HARRIS Mountaintop
Return to Al Sands or Bob Murphy for Approval

Figure 7.4. Zero scrap action report.

Support

Technical, managerial, and logistical support is vital for a program such as Zero Scrap. Teams may require training in problem-solving techniques or may need help from engineering. The organization must supply this support willingly to make the program work. Engineers and technicians need to work with the teams on the production floor or during team meetings. While the frontline production workers are closest to the job, the engineers can provide technical insight and guidance. This is especially important on off-shifts where availability of technical personnel is usually lower.

Mountaintop used the following tactic to show support and encourage the teams. The plant formed a group that included members of the steering committee, staff, process and product engineers, and manufacturing leaders. This group interviewed every team on every shift. The goal was to find out what the plant could do to support the team's scrap reduction effort.

The group did not seek to solve the problems for the team; it sought to elicit the reasons for scrap in each area. The group wanted to learn what technical, manufacturing, or capital support the teams needed to succeed. The group assessed each team's needs in its respective area.

The group collected information from all three shifts and wrote a report. The report included a recommendation to Harris' management for capital and expense money. Most of this money was for small tools or repairs that would have an immediate payback in scrap reduction. With the payback so dramatic, Mountaintop had no problem justifying the outlays in the next capital approval cycle. This tangible support strongly affirmed management's commitment to goal-oriented team processes.

Results

Within a year, Mountaintop teams implemented more than 100 scrap reduction actions. Most of the plant's 65 teams, and all three shifts, participated. The Pareto Principle applied here, too, with one particularly active team being responsible for many improvements. The projects cost about $687 thousand for small tools and expense items. Savings came to more than $6 million, or a 773 percent return in one year.

Meanwhile, Figure 7.5 shows the reduction in scrap wafers at probe. Remember that this is particularly important when the process is working at capacity. The revenue gains from avoiding zero yield wafers are pure profit (see Figures 7.5 and 7.6).

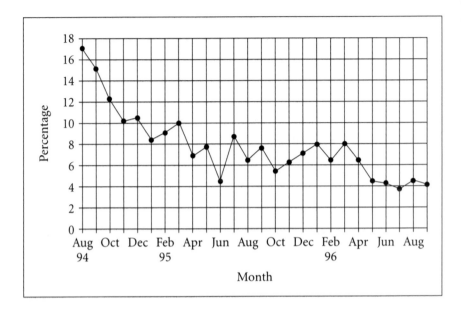

Figure 7.5. Zero-yielding wafers at final test.

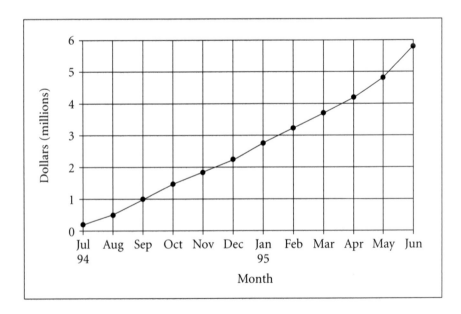

Figure 7.6. Cumulative revenue gain from avoiding zero yield wafers.

The program had an extra benefit: It helped the teams understand and appreciate wafer revenue value. It also encouraged some teams to request extra training in problem-solving methods.

Conclusion

The results of Mountaintop's zero scrap program show that self-directed teams of (primarily) hourly workers can achieve outstanding results. Management must show commitment to the program and must provide support and guidance. If management does its part, workers will use their intimate familiarity with the process to deliver results.

.

CHAPTER 8

Total Productive Maintenance at Harris Semiconductor Mountaintop

Michael Caravaggio

In June 1989, Harris Semiconductor had recently purchased the Mountaintop plant. The company's equipment maintenance managers met at the facility in Research Triangle Park, North Carolina. The meeting's purpose was to introduce total productive maintenance (TPM) to the semiconductor manufacturing plants. We sometimes refer to it as *tool productivity management*, since its goal is to increase the tool's productivity effectiveness.

The equipment managers had to overcome several paradigms to get small groups to buy into the activity. Some of these were the following:

- "Equipment maintenance technicians must be under the control of a central equipment maintenance organization."
- "Process technicians are responsible for solving process-related problems."
- "Operators operate the tools and cannot solve equipment or process problems."
- "This is a union shop. The union will never buy in to the TPM concept."

The following sections will show how the site overcame these paradigms and implemented TPM.

Application of Cross-Functional Work Teams

The site reorganized equipment maintenance technicians and process technicians under manufacturing. That is, the site integrated technical support into small area teams with the operators. The organizational structure no longer separated the technicians from the operators. This organizational change illustrates W. Edward Deming's advice to break down barriers between departments. The union did not object, and the union environment was never a factor.

Mountaintop's upper management showed commitment to the TPM effort and provided a critical ingredient: patience. We must remember that it takes time to train and develop people in the TPM methodology. Management must allow employees time to carry out the improvements. The key factors are patience, persistence, and determination.

Something was, however, missing. Discussions focused on the tools, but not on the process. We needed process engineers, and we added them to the teams. Chapter 11 will show how the site extended this methodology by introducing integrated yield management (IYM). TPM reduces equipment downtime and equipment-related defects, while IYM addresses process-related defects.

Getting Started

Here are the prerequisites for TPM.

- Upper management must provide support and patience.

- There must be a TPM champion (a person well versed in TPM methodology).

- Small work groups make TPM work. Therefore, we must develop teams that include operators, maintenance and process technicians, engineers, and equipment repairers who will take on tool and process responsibility. Chapter 5 describes teaming and its role in the Mountaintop culture.

- Software support must be available.

- TPM must focus on the bottleneck operations or constraints. TPM supports synchronous flow manufacturing (SFM), which is another key Mountaintop program.

  —The entire facility uses TPM methodology, but the major focus is on the bottleneck tools or constraints. Chapter 9 describes SFM and the Theory of Constraints (TOC). The constraint is the operation with the least capacity, which limits the factory's capacity. To increase the factory's capacity, we must make the constraint as efficient as possible (Murphy, Saxena, and Levinson 1996). We will use the photo mask aligners as an example in this chapter.

Questions About TPM

What Is It? TPM is a process for achieving zero defects and zero breakdowns. It reduces equipment-related scrap, rework, and defects, and it improves equipment availability.

Who Should Do It? TPM is a cross-functional activity that requires support from operators, engineers, managers, equipment vendors, and other support personnel. Recall that Mountaintop integrated process and equipment technicians into the self-directed work teams. This removed some barriers between operators and their technical support. We will later see how operators can detect abnormalities and perform routine maintenance tasks.

Effectiveness Measurements. TPM redefines equipment performance. We formerly looked only at uptime and downtime. TPM recognizes that losses occur even while the equipment is "up." TPM uses five loss measurements to track equipment performance:

1. *Failure losses* are "hard downtime," which includes waiting time, repair time, and qualification time. These reduce the tool's availability, or fraction of time when it is available to do work. Preventive maintenance also reduces availability, but not as much as the trouble it prevents.

2. *Setup losses* are time spent changing over the equipment from one product or process to another. Setups also reduce a tool's availability.

3. *Minor stoppages and idle time* occur when the equipment stops running for short intervals waiting for operator intervention. Idle time reduces the tool's operating efficiency.

4. *Reduced speed losses* occur when the actual operating speed is less than the ideal or designed speed or the tool operates with a partial load. These losses reduce the tool's rate efficiency.

5. *Quality losses* occur when rework and scrap reduce the rate of quality, or the percentage of good output from the workstation.

Note that there is one more loss that we choose to ignore: reduced yield caused by the initial setup and final tool stability. TPM activities identify the major losses, discover their causes, and remove these causes to prevent these losses.

Elements of TPM

Table 8.1 shows the five major elements of TPM (Levinson and Tumbelty 1997, Table 2.2; see also Shirose 1992, 11.)

Overall Equipment Effectiveness

Overall equipment effectiveness (OEE) measures the effectiveness of TPM efforts. It is a mistake, however, to apply OEE to the entire factory (Murphy, Saxena, and Levinson 1996). Mountaintop uses it to focus on critical operations—the constraints or bottlenecks. OEE is the product of four efficiencies that represent the major losses: availability, operating efficiency, rate efficiency, and rate of quality.

1. A workstation's *availability* is the percentage of time that it is available to do work. If the factory works three shifts (24 hours) and the station is usable 18 hours a day, its availability is 75 percent. Failure losses and setup losses reduce the tool's availability.

   —Failure losses are never desirable, since it costs time and money to fix them. They may also reflect problems that can degrade product quality.

   —Preventive maintenance reduces availability, but it averts failure losses that are even more costly.

2. *Operating efficiency* is the portion of available time that the workstation uses. If the machine is available 20 hours a day and works 16 hours a day, its operating efficiency is 80 percent. Minor stoppages and idle time reduce the operating efficiency.

   —Unless the workstation is a constraint, idle time is acceptable and even desirable.

Table 8.1. Elements of TPM.

1. Activities that improve equipment effectiveness	Reduction of 1. Breakdowns 2. Setups and adjustments 3. Idling and minor stoppages 4. Speed losses 5. Defects, rework, and scrap 6. Startup losses (losses due to a setup change or machine adjustment)
2. Autonomous maintenance by manufacturing operators	Manufacturing operators perform routine maintenance activities such as cleaning, inspection, and lubrication. They also learn to recognize and respond to abnormal conditions.
3. Planned maintenance	Planned maintenance is similar to an automobile's maintenance schedule. For example, the owner's manual may recommend an oil change every 3000 miles. There is also a schedule for transmission service, wheel bearing cleaning and repacking, and so on.
4. Improvement of operation and maintenance skills through training	Autonomous maintenance means production workers perform routine maintenance tasks. Workers also learn to identify abnormalities, preferably before they cause breakdowns (Shirose 1992, 1–4). Maintenance workers and machine attendants learn advanced skills and techniques. Nakajima (1989, 330) compares the operator to an automobile driver. The driver performs routine tasks like checking fluid levels and tire pressure. The driver can add oil or transmission fluid, or put air in the tires. An auto mechanic (the maintenance worker) must, however, perform complex repairs or maintenance tasks. For example, most drivers have neither the training nor the equipment to do a tuneup.
5. Design for maintenance prevention (MP) and early equipment management	This usually happens during equipment design and does not involve manufacturing operators. The idea is to design new equipment with reduced maintenance needs. Another goal of MP is to make preventive maintenance and repair easier. For example, is it easy or difficult to reach your car's oil dipstick? How hard is it to replace a headlamp or turn signal? Other aspects of MP include design for ease of operation, quality, and safety.

3. The *rate efficiency* is the ratio of the actual production rate to the theoretical rate. If the machine's theoretical capacity is 100 pieces an hour and it makes 70, the rate efficiency is 70 percent. Equipment wearout, operation at less than full speed, control (nonproduct) workpieces, and partial loads reduce the rate efficiency.

—Unless the workstation is a constraint, we do not need to push the rate efficiency. Partial loads, nonproduct workpieces, and slow operation are acceptable. If low efficiency is symptomatic of wearout, however, we should correct it.

4. The *rate of quality* is the fraction of the tool's output that is good. If the tool makes 1000 pieces and 900 are good, the rate of quality is 90 percent. We always want a 100 percent rate of quality, since rework and scrap are never desirable.

OEE is the product of these four efficiencies.

$$\text{OEE} = \text{Availability} \times \text{Operating efficiency} \times \text{Rate efficiency} \times \text{Rate of quality}$$

or

$$\text{OEE} = \frac{\text{Uptime}}{\text{Total time}} \times \frac{\text{Operating time}}{\text{Uptime}} \times \frac{\text{Total pieces}}{\text{Theoretical throughput}} \times \frac{\text{Good pieces}}{\text{Total pieces}}$$

which simplifies to

$$\frac{\text{Operating time}}{\text{Total time}} \times \frac{\text{Good pieces}}{\text{Theoretical throughput}}$$

TPM and OEE

Table 8.1 listed six problems that TPM suppresses. They affect OEE as follows.

1. Breakdowns reduce the availability.

2. Setup and adjustment losses also reduce availability. Shirose (1992, 34) says that 70 to 80 percent of adjustments may be avoidable.

These adjustments result from accumulation of precision errors or from inconsistent measurement methods and standards.

—Note the interaction with statistical process control (SPC). SPC warns production personnel when the process average or variation changes. TPM may reduce the incidence of out-of-control situations.

3. Idling and minor stoppages reduce the operating efficiency. Since they are usually easy to correct, factory personnel often don't look for a permanent correction. When it is easy to work around a problem, the problem eventually becomes an accepted part of the job.

4. Speed losses reduce the rate efficiency.

5. Defects, rework, and scrap lower the rate of quality.

6. Startup and yield losses can affect different efficiencies, depending on their effect. Startups and shutdowns are often problems in chemical process industries, where plant design assumes stable operating conditions. It often takes time for the chemical plant to achieve stable operation, and conversions and yields may be lower until stabilization is achieved.

OEE: Example

Here is a sample OEE calculation. A machine is available 18 hours a day. It operates 15 hours and makes 1200 pieces. Of these, 1080 are good. The machine's theoretical throughput is 100 pieces an hour.

The availability is (18 hours ÷ 24 hours), or 75 percent.

The operating efficiency is (15 hours ÷ 18 hours), or 83.33 percent.

The rate efficiency is (1200 pieces ÷ 1500 pieces), or 80 percent.

The rate of quality is (1080 pieces ÷ 1200 pieces), or 90 percent.

75% × 80% × 80% × 90% = 45 percent.

$$\text{OEE} = \frac{18\,\text{hours}}{24\,\text{hours}} \times \frac{15\,\text{hours}}{18\,\text{hours}} \times \frac{1200\,\text{pieces}}{1500\,\text{pieces}} \times \frac{1080\,\text{pieces}}{1200\,\text{pieces}} = 0.45, \text{or } 45\%$$

Alternately, one machine with a 90 percent OEE could do the work of two 45 percent machines. In practice, however, it is difficult to achieve 90 percent efficiency.

OEE and SFM/TOC

OEE is not a stand-alone activity. It must support synchronous flow manufacturing and the Theory of Constraints. It is a common mistake to measure every operation's OEE and exhort production workers to improve efficiencies. Figure 8.1 shows what happens.

Dogmatic focus on local OEEs can also prevent a factory from improving its throughput. This may sound strange. If we improve all the efficiencies, don't we get more throughput? The constraint, not the local efficiencies, dictates how much product the factory can make (see Figure 8.2). Chapter 9, on SFM, will discuss how idle time and partial loads are acceptable at nonconstraints. That is, fractional operating and rate efficiencies are acceptable and even desirable at these operations. Fractional availability (meaning downtime) and fractional rates of quality are undesirable anywhere. Downtime for preventive maintenance is acceptable, since an hour of preventive maintenance may prevent a shift (or more) of involuntary downtime. However, efforts to improve availability must still focus on the constraint. Downtime at a nonconstraint is recoverable; downtime at the bottleneck isn't.

OEE and Cost Accounting

The cost accounting system will not recognize a problem with inventory, because inventory is a current asset. Inventory does, however, tie up cash. Suppose that the company starts with $100,000 in cash and $40,000 in current liabilities. The company then turns $75,000 of the cash into inventory. Table 8.2 shows what happens.

A current asset is something we can, in theory, quickly turn into cash. Current assets include inventory, work in process, accounts receivable, and, of course, negotiable securities and cash. We cannot, however, pay bills with inventory. Some financial institutions buy receivables, but at a discount, and banks may accept them as collateral. Only cash and negotiable securities are dependable means of paying immediate obligations. Liquid or "quick" assets include only cash and negotiable securities.

A 2.0 or better current ratio is usually healthy, as is a 1.0 quick ratio. After the factory has turned its cash into inventory, the current ratio is still 2.5. The quick ratio, however, is down to 0.625. *Poor cash flow can*

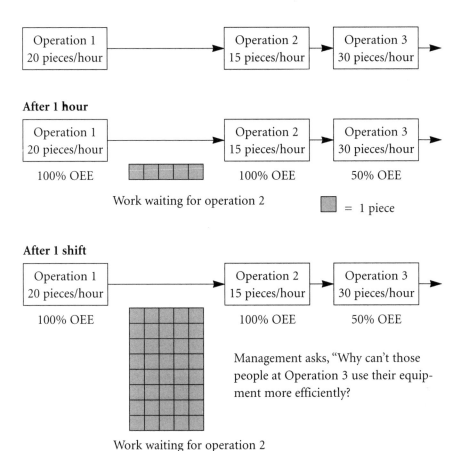

After 1 hour

100% OEE — Work waiting for operation 2 — 100% OEE — 50% OEE

▢ = 1 piece

After 1 shift

100% OEE — 100% OEE — 50% OEE

Management asks, "Why can't those people at Operation 3 use their equipment more efficiently?

Work waiting for operation 2

After 1 year

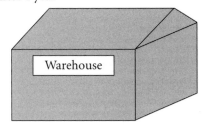

Management asks, "Why can't those people at Operation 3 use their equipment more efficiently? And why are we renting a warehouse to store inventory?

Work waiting for operation 2

Figure 8.1. Misuse of OEE to measure performance.

Figure 8.2. Operating efficiency paradigms and SFM/TOC.

bankrupt even a profitable company. If we misuse OEE by applying it to every operation, we can bankrupt ourselves very efficiently. Readers of Goldratt and Cox's *The Goal* (1992) will appreciate this situation.

Chapter 9, on SFM/TOC, shows how Mountaintop manages its production and inventory.

Table 8.2. Financial impact of misusing OEE.

	Before	After
Cash	$100,000	$25,000
Inventory	None	$75,000
Current or short-term liabilities	$40,000	$40,000
Current ratio (divide current assets by current liabilities)	$\dfrac{\$100,000}{\$40,000} = 2.5$	$\dfrac{\$25,000 + \$75,000}{\$40,000} = 2.5$
Quick, or acid-test ratio (divide liquid assets by current liabilities)	$\dfrac{\$100,000}{\$40,000} = 2.5$	$\dfrac{\$25,000}{\$40,000} = 0.625$

Automated Data Collection and OEE Calculations

A factory earns money by making product, not by collecting information. Information is useful only if it helps the factory make product. Data collection should be automatic and should require little or no operator intervention. The information system must collect accurate data, and should avoid using the operators' time. Mountaintop's automatic data collection system has three parts.

1. The lab tech control data collection system extracts the tool status. Mountaintop uses the status codes shown in Table 8.3. The tool has a switch that returns an analog signal.

2. Workstream is a logistical management information system. It extracts product throughput, scrap, and rework counts.

3. The Ilmtp database is the storage and retrieval area for the data. This database supplies the raw data for the OEE calculation in the spreadsheet (Aligner Fab 5).

Figure 8.3 shows a histogram of the reasons that the workstation is not operating.

Table 8.4 and Figure 8.3 reflect several months' work by the photolithography teams to nearly triple the OEE. The original efficiency was 26

Table 8.3. Machine status codes (photolithography).

Code	Status
0	Machine available (machine can run)
1	First pattern setup (setup prior to first production wafer; first pattern)
2	Preventive maintenance focus distortion, lamp, uniformity)
3	Waiting for parts (waiting for vendor-supplied parts)
4	Facilities problem (loss of nitrogen, humidity control, temperature, high-pressure air)
5	Meetings (machine idle, operators at meeting)
6	Hard down (tool is down for unscheduled repair or breakdown)
7	Waiting for technical help (waiting for technician, equipment repairer, engineer)
8	No product (no product at machine)
9	Minor stoppages (intermittent errors; time is automatically calculated when the yellow light is on, but can be separately keyed in)

percent, and it is now close to 70 percent. The area team focused on and removed the root causes of the inefficiencies, thereby achieving world-class results. For the photolithography equipment, the major root causes were the following:

1. Misalignment resulted from an infinite setting for the runout selection switch. This disabled the alarm for excessive runout. The setting was readjusted for a runout of four. This reduced misalignment reworks and increased the rate of quality. (In the semiconductor industry, misalignment means that successive layers of the microelectronic device do not line up. Imagine two metal pieces that are to be fastened with screws, but the holes do not line up. Misalignment is similar.)

2. Manual operation caused a loss of rate efficiency. The problem was a malfunction of the kicker arms. A modification to the aligners removed the need for the kicker arms and increased the rate efficiency.

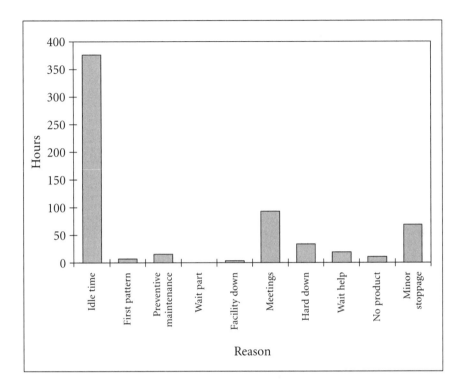

Figure 8.3. Histogram of nonoperating time.

3. Streamlined work flow to the aligners reduced the "waiting for work" status and increased the operating efficiency. ("Starving the constraint" means the constraint operation cannot run because there is no work.)

Hazards of Using the Wrong Metric. The just-mentioned tool inefficiencies resulted from losses that occurred while the tool was available to run. Under the old metric of uptime versus downtime, no one would have looked for root causes for inefficiencies. The availability was 92 percent, and everyone would have assumed this to be satisfactory.

Focus OEE on the Constraints. Mountaintop focused TPM and OEE on the constraint operations. These were the metal evaporators in Fab 4, photolithography equipment (aligners) in Fab 5, and sputtering equipment

Table 8.4. Spreadsheet for OEE calculation.

Idle time	380.15	Available time	2643.31
First pattern	2.81	Idle time	487.06
Preventive maintenance	12.83	Run time	2156.25
Wait part	0		
Facilities down	1.91	Availability	95.10%
Meetings	95.74	Operating efficiency	81.57%
Hard down	36.44	Rate efficiency	93.99%
Waiting for help	18.31	Rate of quality	95.00%
No product	11.17		
Minor stoppage	63.89	OEE	69.27%
Run time	2156.25		
Total time	2779.5		
Available time = Total time minus first pattern, preventive maintenance, waiting for parts, facilities down, waiting for help, hard down, and minor stoppages. Run time = Available time minus idle time, meetings, no product.			

in Fab 6. (As previously explained, the fab number refers to the wafer size; for example, Fab 4 processes four-inch diameter silicon wafers. The new Fab 8 production line handles eight-inch wafers.)

5S-CANDO

5S-CANDO is a set of activities that support TPM. The five Ss stand for five Japanese words: *seiri, seiton, seiso, shitsuke,* and *seiketsu.* Their rough English equivalents are *clearing up, arranging, neatness, discipline,* and *ongoing improvement:* CANDO.

Clearing Up. This step removes unnecessary items from the workplace. Classify every piece of material and every tool in the work area by frequency of use (daily, weekly, monthly, or never). Items that people use every day should be at the workstations. Items that people need weekly or monthly should be available, but do not need to be at the stations. Remove items that no one ever uses.

People can use red tags to identify items they believe to be nonessential. The tag is a form of cross-shift communication. If no one claims the item, it is safe to remove it after a predetermined time.

Arrangement. Standardize responsibilities and procedures. Create a common language of visual signals to reduce mistakes and to make the work rules transparent. Make deviations from standards obvious and mistakes self-regulating.

A key phrase is, "A place for everything and everything in its place." A common example in machine shops and garages is a wall rack for wrenches. Each wrench has its own place in the rack. If people return the wrenches to their places, no one has to spend time searching for the tools. Another example is a socket wrench set with a labeled position for each socket. These are examples of visual controls (see Figure 8.4). Extend this thinking to the rest of the workplace.

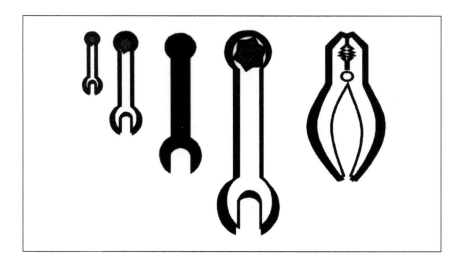

Figure 8.4. Tool rack as visual control.

Neatness. Clean the equipment, the tools, and the entire workplace. A spotless workplace will often reveal abnormalities that would otherwise remain hidden. A leak over a clean floor will quickly become apparent, while it might go unnoticed on a dirty floor. Use cleaning as a form of inspection to expose abnormalities. Also, cleaning removes dirt and debris that can interfere with the product or equipment.

Discipline. Discipline means standardizing operations and activities, then following the standards. Employees help create rules, procedures, and checklists. This reinforces commitment and support for the CANDO program. Management also must show commitment by providing resources and support.

Ongoing Improvement. Set things in order, set limits, and share information. Generate and adopt new and innovative ideas and aggressive goals. Do not accept abnormalities or let them become part of the daily routine.

Summary: 5S-CANDO Principles. Rudyard Kipling's poem *The 'Eathen* is an excellent summary of 5S-CANDO and its goals (*abby-nay, kul,* and *hazar-ho* mean "not now," "tomorrow," and "wait a bit"). Kipling recognized the role of discipline—of doing things "just so"—in the British Army's superiority over most others.

> *Keep away from dirtiness—keep away from mess*
> *Don't get into doin' things rather-more-or-less!*
> *Let's ha' done with abby-nay, kul, and hazar-ho;*
> *Mind you keep your rifle an' yourself jus' so!*
>
> —Rudyard Kipling, *The 'Eathen*

1. Keep necessary tools and materials at workstations, but don't keep unnecessary or extra ones there.
2. Keep tools and work areas clean.
3. Use visual controls.
4. Pride in the workplace and operation is evident.
5. Standard procedures are in place and are easy to understand.

Root Out Inefficiencies and Abnormalities. Acceptance or assimilation of apparently minor inefficiencies into a job can undermine a factory's competitiveness. Workers may accept chronic, apparently minor problems that they can work around. A simple example is a machine that stops working, but starts again if a worker kicks it in the right place. The workers become used to kicking the machine several times a shift, and the inefficiency becomes part of the job. (If someone took the time to open the cover and fix the loose connection, the problem would go away.) Problems such as these are like tapeworms: They don't cause an acute illness, but they absorb resources.

"Hidden plant" refers to the capital investment and overhead that make up for the inefficiencies. Suppose that 20 percent of everything a factory makes is rework. This plant must have 25 percent more equipment, floor space, people, and overhead to make up for the rework. If the accounting system includes allowances for scrap and rework, people accept these as normal parts of the operation. Juran and Gryna (1988, 4.11–4.12, 22.5–22.6) discuss hidden costs. There is never any incentive to correct them or look for their root causes.

The Prussian general Carl von Clausewitz (1976, book 1, ch. 7) referred to such problems as *friction*. "Friction, as we choose to call it, is the force that makes the apparently easy so difficult . . . Countless minor incidents— the kind you can never really foresee—combine to lower the general level of performance, so that one always falls far short of the intended goal." Tom Peters echoes this principle in *Thriving on Chaos:* "The accumulation of little items, each too 'trivial' to trouble the boss with, is a prime cause of miss-the-market delays" (Peters 1987, 323).

Friction is insidious and treacherous because people can "work around it." The 5S-CANDO philosophy says that friction is not acceptable. Identify its root causes, eradicate it, and don't allow it to return.

5S-CANDO at Mountaintop

The Mountaintop facility started the implementation of the CANDO program in June 1993. The entire workforce, including operators, engineers, and the support functions, were trained on CANDO. The Mountaintop teams were receptive to the program, and they used the techniques listed in Table 8.5.

Table 8.5. Mountaintop's techniques for supporting CANDO.

Technique	Description
Classification list	Classify every piece of material and every tool in the work area by frequency of use (daily, weekly, monthly, or never). Get rid of extraneous items.
Organization lists	Use the classification list to assign a place for every piece of material and every tool. Items that people use every day belong in the work area. Items used weekly should go in separate material/tool areas, and items used monthly should go in more remote areas. Use this list to organize the work area and to develop standard places for every item.
Fixed point photography	Photograph the work area periodically from the same position. Display the photos on a chart in a central location. Develop an action plan to improve the work area, and post it below the picture. The series of pictures lets workers see the progress and reinforces the effectiveness of 5S-CANDO. Quality tools are always more effective if the people who use them can see their results.
CANDO calendar	Post a calendar showing when regular tasks need to be done.

In summary, the Mountaintop teams received the information and training necessary to perform autonomous maintenance on their respective tools. This promoted pride in the workplace and empowered the teams to take responsibility for the process and equipment.

TPM, IYM, and TOC are synergistic and mutually supporting programs. They are Mountaintop's nucleus for improving throughput and yield.

Visual Control Systems

A visual control system is an important tool for combating friction. Visual controls identify waste, abnormalities, or departures from standards. They are easy to use even by people who don't know much about the production area.

Visual controls support 5S-CANDO by excluding unnecessary items from the work area. They support the concept, "A place for everything and everything in its place." A visual control system has five aspects.

1. Communication: Written communications are easily accessible.

2. Visibility: Communication with pictures and signs.

3. Consistency: Every activity uses the same conventions.

4. Detection: There are alarms and warnings when abnormalities occur.

5. Fail-safing: These activities prevent abnormalities and mistakes.

Effectiveness of Communications

Visual and pictorial communications are often more effective than written communication. With Indo-European phonetic languages, the reader must translate abstract letters into mental images and ideas.* A picture conveys the information without the need for translation. Political cartoons are effective because they convey an idea in a few seconds. An editorial may take a couple of minutes to do this—that is, if the viewer's attention span is long enough to read it.

Verbal instructions have many deficiencies and are easy to misunderstand. Gilbert and Sullivan's *The Pirates of Penzance* gives a humorous example. A father instructs a nursemaid to apprentice his son to a ship's pilot. The nursemaid later says, "Mistaking my instructions which within my brain did gyrate/ I took and bound this promising boy apprentice to a *pirate.*"

Does this happen only in Gilbert and Sullivan comedies? Unclear directives during the Crimean War caused the disastrous Charge of the Light Brigade: The British cavalry attacked the wrong guns. During the First World War, unclear signals undermined the Royal Navy's performance in three battles. At Dogger Bank (1915), unclear signals allowed the

---

*Chinese and Japanese *kanji,* or ideographs, represent ideas. For example, the symbol for "forest" actually looks like a group of trees. We are not intimately familiar with these languages, but it's very possible that they facilitate thought processes that are unfamiliar to Westerners.

German battle cruisers to escape. At Jutland (1916), poor signaling helped lose the battle cruiser *Indefatigable* with almost all hands. The same flag-lieutenant was responsible for these errors, and he was not even a signal officer. Admiral Beatty himself said, "He lost three battles for me," but he did not dismiss or reassign him (Regan 1987).

You may wonder why, in war movies that involve radio communications, people talk funny. They'll say, "Rendezvous at Alpha Mike Niner" or something similar. The term "Charlie" for a Viet Cong guerrilla came from "Victor Charlie," or VC. Military organizations have long known the need for clear communications, and this is a form of mistake-proofing. "D" sounds like "t" and "b" sounds like "v," but one cannot mistake "delta" for "tango" or "bravo" for "Victor." "M" and "n" sound alike, but "Mike" sounds nothing like "November."

Written communications leave less chance of being misunderstood, but visual controls are even easier to understand.

Examples of Visual Controls

Visual controls in everyday life include traffic control devices such as stop signs and traffic lights. Redundancies ensure that motorists will not misunderstand the devices. Stop signs are always red octagons, and no other sign is octagonal. This is an example of consistency. A traffic light always has red on top and green at the bottom, in case the driver is color-blind. We take these conventions for granted. A white rectangular "stop" sign at an intersection would cause a lot of confusion, and probably cause collisions. Most people probably wouldn't stop for it. Many might stop for a red octagonal "no left turn" sign at an intersection, which would probably cause a lot of rear-end collisions.

The tool rack in Figure 8.4 is an example of a visual control in a factory. Each tool has its place, and it is easy to see if a tool is missing. It is also easy to see where one should return a tool.

Autonomous Maintenance

The mission of autonomous maintenance (AM) is to recognize, restore, and prevent deterioration. Operators can perform routine maintenance tasks and even improve the operation (see Figure 8.5).

Figure 8.5. Traditional versus autonomous maintenance.

Routine equipment checks reveal deterioration that could lead to breakdowns. Routine cleaning will often reveal defects and other problems.

Operators can restore equipment by fixing abnormalities, replacing simple parts, and helping with emergency repairs. They can also help maintenance workers and engineers with improvements.

Production workers can prevent deterioration by maintaining basic equipment conditions and setting up and operating equipment properly.

They can collect and record data on abnormalities and help engineers and maintenance workers improve methods and equipment.

AM improves morale and motivation. It gives production workers broader responsibility for the operation and its performance. AM upgrades the maintenance workers' role by requiring them to focus on high-skill tasks.

Responsibilities for AM

Harris Semiconductor's AM program gives operators responsibility for the following tasks.

1. Operators set up and operate the equipment, and perform engineering qualifications. A certification program ensures that they have the necessary training and experience.

2. Operators perform daily checks and cleanings, and minor repairs and preventive maintenance.

3. Operators identify, measure, correct, and prevent deterioration.

4. Operators help maintenance workers and engineers with improvements.

5. Operators maintain optimal equipment conditions.

Benefits of AM

Autonomous maintenance provides the following benefits.

1. The shop floor is clean and well-organized. 5S-CANDO also supports cleanliness and organization.

2. Everyone can see evidence of improvements.

3. AM promotes morale and motivation.

4. Equipment is clean and operates under optimum conditions.

5. Standards are present, consistent, and universally understood.

6. Production and maintenance workers work together with engineers on improvements.

7. Visual controls support the program.

8. Abnormalities and problems are obvious.

Implementation of AM

There are 14 steps for implementing autonomous maintenance.

1. Form cross-functional small group activity teams.

2. Start 5S-CANDO stage 1: Clearing up.

3. Perform initial cleaning with tags and checklists. Pay attention to equipment abnormalities and problems, and become familiar with the equipment.

4. Get rid of contamination sources and make areas easy to clean.

5. Start visual control system.

6. Perform 5S-CANDO stages 2 and 3: Organizing and cleaning.

7. Set up maintenance standards for cleaning and inspection.

8. Assess and improve engineering and maintenance skills.

9. Increase knowledge and understanding of the equipment. Learn the best performance conditions and learn to diagnose problems and prevent deterioration.

10. Perform 5S-CANDO stage 4: Standardization.

11. Develop operator ownership of the equipment. Operators learn to recognize and correct abnormalities.

12. Perform 5S-CANDO stage 5: Discipline. Refine the visual control system.

13. Skills transfer and operator training.

14. Autonomous maintenance becomes the factory's way of doing business. Continuous improvement becomes part of "the way we do things around here."

Completion of these steps leads to the following levels of AM capability.

Level	Steps Completed	Capability
4	All	Operators can repair equipment
3	11–12	Operators understand the relationship between equipment conditions and quality.
2	8–10	Operators learn about equipment functions and structure.
1	1–7	Operators can recognize abnormalities and improve the equipment.

Implementation Activities

The following programs and activities support autonomous maintenance.

1. Small group activities

2. One point lessons (basic information, problems, improvement)

3. 5S-CANDO activities

4. Visual controls

5. Skill enhancement training for operators, maintenance workers, and engineers

6. Activity boards

Benefits of TPM

Mountaintop's TPM program has led to continuous tool improvement, which yields the following benefits.

1. Maximizes tool utilization

2. Increases productivity

3. Provides a focal point for team activities

4. Provides information for capital investment planning

5. Reduces scrap and rework

Examples

Here are two examples of Mountaintop's application of TPM.

Lam Etcher. The Lam AutoEtcher 690 is an aluminum plasma etch system. The manufacturer is Lam Research, Inc. The etcher is a fully automated single wafer plasma etching system for etching aluminum on a silicon substrate. This operation follows the metal sputtering and photolithography processes (see Figure 8.6).

The Lam Etcher team's OEE started at 35 percent. Several months of focused TPM efforts improved the OEE to 58 percent. A Pareto chart showed the major loss sources, and the team focused its efforts accordingly. Here is the analysis.

1. Radio frequency problems were the major loss, and a modification to the phase lock loop radio frequency generator got rid of them.

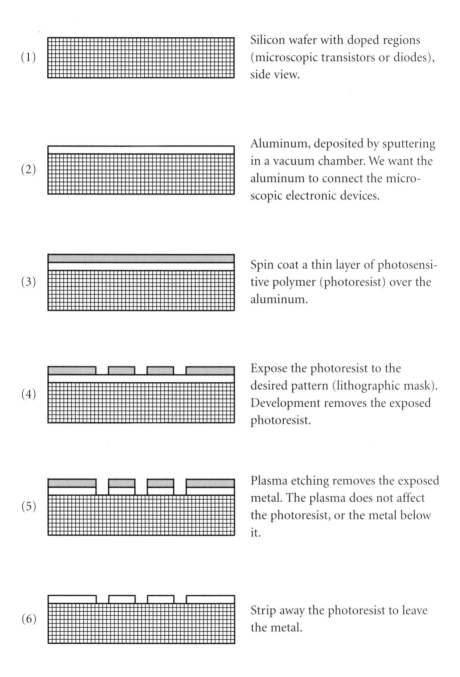

(1) Silicon wafer with doped regions (microscopic transistors or diodes), side view.

(2) Aluminum, deposited by sputtering in a vacuum chamber. We want the aluminum to connect the microscopic electronic devices.

(3) Spin coat a thin layer of photosensitive polymer (photoresist) over the aluminum.

(4) Expose the photoresist to the desired pattern (lithographic mask). Development removes the exposed photoresist.

(5) Plasma etching removes the exposed metal. The plasma does not affect the photoresist, or the metal below it.

(6) Strip away the photoresist to leave the metal.

Figure 8.6. Role of plasma etching in semiconductor fabrication.

2. Pressure problems came from intermittent point-of-use valve failures.

3. Cleanings and high leak rates reduced the tool availability. Investigation showed their sources as edge bead removal (EBR) from the photolithography process, and the process's use of $CHCl_3$ (chloroform). EBR is the removal of the photoresist bead that forms around the wafer's edge during spin coating.

4. The team installed a low-level warning system before a system shutdown. This ended the cooling problems.

5. Rebuilding of the gas isolation valve fixed gas problems, especially excessive consumption of CH_4 (methane).

6. O-ring deterioration apparently came from Buna (polybutadiene) o-rings that infiltrated into the stock system. Buna was purged from the inventory.

Ion Implanters. Ion implanters use ion beams to drive dopant atoms into silicon wafers (see Figures 8.7 and 8.8). Dopants change the material's electrical properties and allow creation of transistors and diodes. Mountaintop's first TPM effort was in the ion implanter area. This was a throughput constraint, with capital cost exceeding $1 million per tool, and the tools were very maintenance intensive. The initial OEE was about 23 percent. Several months of a strong TPM effort improved the OEE to a respectable 48 percent. This doubled the throughput, or halved the capital investment and maintenance. Some implant TPM improvements were the following:

1. Reducing the use of control (nonproduct test) wafers to increase capacity and reduce variable cost.

2. Upgrading the software to include vacuum interlocks. This improves implant quality.

3. Calibrating the mass spectrometer to read the correct atomic mass unit (AMU) reduced errors and setup time.

4. Replacing corroded air hoses during planned downtime periods. This keeps them from breaking and causing unplanned downtime.

5. Changing to vapor blasting instead of bead blasting of source parts to reduce foreline clogging and excessive pump down time.

Figure 8.7. Ion implanter (open, with wafers).

Figure 8.8. Ion implanter (closed).

6. Changing from 15 percent to 7 percent arsenic in hydrogen to reduce foreline clogging. Arsenic is a common dopant, and it comes as arsine (AsH_3) gas.

7. Overhaul the vacuum systems to improve the rate efficiency.

8. Upgrade tools to 120 kev (120,000 electron volts) to increase versatility and reduce idle time.

Summary

In summary, TPM can be very successful when the teams use it. TPM, IYM, and TOC/SFM are synergistic and mutually supporting programs for improving throughput and yield. TPM's success also depends on a good tool calibration system.

Synchronous Flow Manufacturing

Robert E. Murphy Jr. and Puneet Saxena

Synchronous flow manufacturing (SFM) is a logical and systematic approach for moving material through production operations. Work should move quickly, smoothly, and in concert with market demand. This production management approach has gained steady momentum since Eliyahu M. Goldratt and Jeff Cox wrote a ground-breaking novel, *The Goal*, in 1984.

People have historically viewed production as an art rather than a science. Frequent expediting, late deliveries, long production lead times, and constant fire-fighting are common in most manufacturing facilities. People expect a manufacturing manager to cope with such symptoms every day. Goldratt's Theory of Constraints (TOC) explains that these symptoms are not independent. They come from a core problem that exists in most manufacturing organizations. Most managers live in the cost world and try to manage organizations by optimizing locally. This chapter will explain the cost model's dangerous deficiencies and break some paradigms. It will then explain the TOC and SFM.

Basic Operational Metrics

Most organizations have a plethora of metrics for the health of their manufacturing operations. They often add new metrics and leave the old ones on the list. The measurements eventually become confusing and sometimes conflicting. They often focus employees' attention on the wrong priorities and cause suboptimization.

TOC defines a simple organizational goal and three straightforward measurements. The goal is simple, but common measurements often displace it: *An organization's goal is to make money now and in the future.*

Theory of Constraints: Performance Metrics

TOC defines three basic operational metrics for measuring performance: throughput, inventory, and operating expense.

1. *Throughput.* Throughput is money that a system generates through sales. It is like gross margin on sales, because it equals the selling price minus the cost of raw materials. Be careful here. Standard cost accounting systems define the gross margin as sales revenues minus direct costs, and direct costs include labor and materials. Direct costs sometimes include direct overhead, too.

Why doesn't TOC define throughput as the sale price minus all the direct costs? TOC deals with engineering and managerial economics, not cost accounting. Companies often get into trouble when they use their cost accounting systems to guide their actions. The cost accounting system should, and legally must, be the basis of tax returns and financial reports. It rarely, however, has much to do with real-world business judgments. That's where engineering economics, and concepts such as marginal and opportunity costs, must prevail.

TOC works with the marginal value of each transaction: the difference between making and selling a unit and *not* making and selling a unit. While accounting systems treat hourly labor as a direct and variable cost, it's really a fixed cost. The factory pays the workers for eight hours a day, whether it uses them or not. Hourly labor becomes a variable cost only when the factory starts paying overtime. Similarly, even direct overhead is not a variable cost. Tom Peters (1987, 587–588) says, "You can't shut the heat off around one machine." Therefore, the true marginal value of a transaction is the sale price minus the raw materials.

2. *Inventory.* Inventory is the money a system invests in items it intends to sell. Traditional accounting systems recognize raw material, work-in-process (WIP), and finished goods as inventory. TOC adds the book value of the plant, property, and equipment.

There is again a similarity between TOC and engineering economics. Traditional cost accounting systems look at materials, WIP, and finished goods. TOC looks at *all* the money the business has invested in the operation. Engineering economics looks at alternate uses and opportunities for money. If we can get a million dollars in throughput from a two million dollar investment, that's great. If it takes a billion dollar investment to do this, the money should be in mutual funds or even in the bank. The basic question is, "How much money do we have to tie up to generate the throughput?"

3. *Operating Expense.* Operating expense is the money the system spends to convert inventory into throughput.

Together, these three measurements can help determine whether any manufacturing system is doing well. Net profit is simply the total throughput minus the total operating expense. Return on investment (ROI) is the ratio of net profit to inventory. At first, it does not look like there is anything radically new about these definitions. A closer look reveals that differences in relative emphasis on these three metrics create the basic distinction between the cost and throughput worlds.

The Cost World

Most production managers live in environments where cost reduction and cost control are the most important goals (see Figure 9.1). Cost reductions improve gross margins on sales, and departments such as marketing, sales, and accounting look at gross margins. The marketing department, for example, wants to show high gross margins on its product lines. Marketing managers adopt or discontinue products because of gross margins. These depend, in turn, on standard costs that use conventional cost accounting principles. Throughput is the next consideration in these environments. Inventory is the last consideration, since conventional wisdom associates inventory with stable output. While cost accounting systems recognize inventory's carrying costs, they also treat it as a current asset.

This is the cost world: Operating expense is first, throughput is second, and inventory is third. The control system measures each department and encourages each department manager to improve locally on these measurements. The words *improve locally* should be frightening; they are almost synonymous with *suboptimization*. Senior managers would never dream of telling department leaders, "Suboptimize the overall process."

A factory's goal is to make money now and in the future. To achieve this, focus on throughput, inventory, and operating expense.

A factory's goals include (1) low costs per piece, which mean higher gross margins on sales, and (2) high efficiencies at every operation.

The Gordian Knot:
"Can't be done"
"Always did it that way"
"It will never work"

Beth Hollock, 1997

Figure 9.1. Traditional metrics versus the Theory of Constraints.

They routinely say, however, "I want higher efficiencies and lower costs from each department."

The Cost Model's Dangerous Deficiencies. Meanwhile, dogmatic adherence to the cost model can endanger the company's financial health. Consider the following example.

- Labor is $10/hour. Each employee can make up to 15 pieces an hour, and 10 employees can produce up to 1200 pieces a day.

- There is a market demand for 800 pieces a day.
- The product uses $5.00 of material per unit.
- The cost accounting system divides $1000 of direct overhead among the day's production.
- Start with $5000 in cash.
- Current liabilities are $4000.

Table 9.1 shows what happens if the company makes extra pieces to "keep the workers busy" and "reduce the cost per piece." Chapter 4, on culture as foundation, discusses how workers at Mountaintop once focused on "making the numbers." The cost per piece goes down, and the gross margin on sales goes up. The marketing and sales people, and the traditional bean counters, love this. Even the current ratio (current assets divided by current liabilities) improves, because inventory is a current asset. There should be a red flag, though, for the quick or "acid test" ratio—this is liquid assets (cash or equivalent) divided by current liabilities. A quick ratio of 1.0 or higher is usually healthy, while less than 1.0 is hazardous.

The company gets the best results by making the 800 pieces that the market can absorb. This choice leaves the company $600 richer, with the highest quick ratio. The 800 piece choice yields $5600 in current assets, as cash. If the factory makes 1200 pieces, it theoretically has $6200 in current assets. The $2600 in inventory, however, is worthless unless the company can sell it. The traditional cost world measurements (cost per piece and gross margin) have guided the company in the wrong direction.

Here's another example of how the cost model can misdirect an organization. The factory has an order for 800 pieces at $8.00 each. Suppose another customer offers $6.00 per piece for up to 400 pieces. The factory wants to know whether to make another 400 pieces. The cost model looks at the $6.50 cost per piece if the factory makes 1200 pieces. "No, don't take that order, we'll lose 50 cents per unit!" What happens if we take the offer? Goldratt and Cox (1992, 311–314) treat this situation.

The factory gets 800 @ $8.00 plus 400 @ $6.00, or $8800. It spends 1200 @ $5.00, or $6000, on material, and pays $800 for labor and $1000 for overhead. The total outlay is $7800, and the factory earns $1000. The company is $1000 richer at the end of the day and has $6000 in current

Table 9.1. Production choices and financial results.

Pieces	600	800	1000	1200	1200*	1350[†]
Labor	$800	$800	$800	$800	$800	$950
Materials	$3,000	$4,000	$5,000	$6,000	$6,000	$6750
Direct overhead	$1,000	$1,000	$1,000	$1,000	$1,000	$1000
Total outlay	$4,800	$5,800	$6,800	$7,800	$7,800	$8700
Cost/piece	$8.00	$7.25	$6.80	$6.50 (best[†])	$6.50	$6.44
Sales	600	800	800	800		
Sales: @$8.00	$4800	$6400	$6400	$6400	$8800*	$9700[†]
Gross margin on sales	$0	$600	$960	$1200 (best[‡])	$1000	$1000
Cash	$5000	$5600 (most)	$4600	$3600	$6000	$6000
Inventory	$0	$0	$1360	$2600	$0	$0
Current assets	$5000	$5600	$5960	$6200	$6000	$6000
Current liabilities	$4000	$4000	$4000	$4000	$4000	$4000
Current ratio	1.25	1.4	1.49	1.55	1.5	1.5
Quick ratio	1.25	1.4 (highest)	1.15	0.9 (danger!)	1.5	1.5

*Another customer offers to buy 400 @ $6.00. 800 @ $8.00 + 400 @ $6.00 = $8800.
[†]Another customer offers to buy 550 @ $6.00. 800 @ $8.00 + 550 @ $6.00 = $9700.
[‡]Cost world standards.

assets (all cash). Had it made only 800 pieces and sold them for $8.00 each, it would have made $600. The cost model said it would lose 50 cents per piece if it sold them for $6.00! What happened?

The cost model treats labor and overhead as part of the cost per piece. It forgets that the factory must pay these costs even if it doesn't make anything. Materials, at $5.00 per piece, are the only variable cost. The marginal value of selling another piece for $6.00 is $6.00 minus $5.00, or a dollar in profit. The *marginal value* of selling 400 such pieces is $400, not a $200 loss as the cost model predicts.

Now suppose the customer offers to take 550 pieces for $6.00 each. The factory must make 1350 pieces, but it can make only 1200 in eight hours. It has to pay the workers for 10 hours of overtime, or $150. This translates into a dollar for each of the 150 extra pieces. Add $5.00 in materials, and the marginal cost of the 150 pieces is $6.00 each. The company gains nothing on this transaction, although the cost per piece ($6.44 versus $6.50) is lower.

This example shows the difference between cost accounting and managerial/engineering economics. The cost world statements belong in the company's tax returns and SEC filings. They have little place in real-world production management.

As manufacturing has become more competitive, people have challenged the traditional rules and assumptions. The throughput world has emerged as the stronger view.

The Throughput World

The throughput world emphasizes the three basic measurements differently. The throughput model recognizes a finite limit for cost reduction and focuses on increasing throughput. It assigns the highest priority to the generation of revenue through sales.

Cost models acknowledge inventory's carrying costs. The throughput model recognizes other problems, too. Eliyahu Goldratt's *The Race* (1996) shows how inventory can affect the quality and price of the product. Inventory also degrades the company's responsiveness to its customers' needs. Higher inventories translate directly into higher production lead times. These, in turn, cause late customer deliveries and frequent rescheduling and expediting on the shop floor. The throughput model ranks inventory after throughput in importance. Operating expense is last, while in the cost world it is first.

The throughput model recognizes a need to control expenses, because they reduce the net profit. This is not, however, the highest priority. The goal is to make money now and in the future. Cost reductions often provide only short-term gains in net profit, but they reduce the organization's ability to generate throughput in the future. Shortsighted cost reductions "kill the goose that lays the golden eggs," and the throughput model avoids them.

This lesson could help investors make money in the stock market. Many investors focus on the bottom line, as reflected by corporate cost accounting systems. "Wall Street ♥ Layoffs" because they reduce short-term expenses. The quarterly statement looks good, and the stock's price goes up. The benefit is temporary: The farmer is displaying the golden egg he got by killing the goose. There won't be more eggs tomorrow or the next day, but many investors don't look that far ahead. Meanwhile, labor is only a small part of the cost in a capital-intensive business. A company cannot "lay off" its capital equipment and its costs. It can only sell the equipment, often at a loss. Remember that the goal of TOC is to make money now *and in the future.* Investors who understand this can outthink investors who don't.

This is the throughput world: Throughput is first, inventory is second, and operating expense is third. The model rates individual departments' performance as they affect the overall system. SFM helps achieve the goals of the throughput world.

Impact on Measurements

The chain analogy shows the difference between the cost and throughput worlds. The chain symbolizes the organization. In the cost world, the chain's *weight* is the primary measurement, while the chain's *strength* is the central performance measurement in the throughput world. The cost model believes that improving the weight of individual links (departments) makes the entire chain better. The throughput world acknowledges that such local improvements do not always add up to provide the global optimum. To improve the chain's strength, one must strengthen the weakest link, or *constraint.* We will see that the constraint is the slowest operation in the process. This is the basis of SFM—or TOC as it applies to production.

Be Careful What You Wish For, You Might Get It

Performance measurements guide organizational behavior. Management asks for performance by defining these measurements, and management always gets what it asks for. What it asks for, however, might not be what it wants. It is a challenge to identify "desirable behavior," but a good start is to define it as behavior that helps the organization achieve its goals. Making money is a primary goal of any business organization. Many organizations, however, often use measurements that force people to act against this goal. Dysfunctional measurements encourage, and even mandate, dysfunctional behavior.

SFM de-emphasizes all extraneous measurements. It focuses only on the three basic ones: throughput, inventory, and operating expense. This simple measurement system guides dramatic changes in traditional attitudes toward production. Local machine and labor efficiencies are no longer important at nonconstraints, and idle time on machines is not always bad. Production workers do not perform unproductive work merely to appear busy. Senior managers no longer rate individual departments on efficiencies. They focus instead on the process' overall efficiency, which depends on the constraint.

This is a paradigm shift, and paradigm shifts are often difficult for organizations. It requires people to abandon preconceptions about how they should run an operation. To sustain this change, the organization's leaders must create and cultivate the proper climate.

Climate for Change

The organization's leaders are responsible for defining the vision and creating the climate for change. Employees will easily revert to old work habits if the new approach falters even a little during the early stages. Skepticism and uncertainty are part of any major change. The leaders at Harris Mountaintop believed in SFM and guided the organization through this uncertain time. Until improvements became visible, management shielded the wafer fabs from dysfunctional metrics, which encouraged people to build inventories merely to keep themselves and the equipment busy. As throughput and production cycle times improved, people bought into the new approach and actively participated in it. It was then easier to sustain the climate for change. Mountaintop's management team has provided TOC education to all its employees, plus steady psychological reinforcement through its visible support of SFM. This also helped create and reinforce cultural change at Mountaintop.

Traditional Production Philosophy

The traditional production philosophy at Harris' Mountaintop plant relied on material starts. The production control system pushed wafers into the line in response to customer demand. To satisfy the customers, the production line had to process the wafers within the estimated cycle time. Management considered the production control department effective if it started the necessary wafers successfully. Any miss in scheduled starts would warrant an investigation, and employees would work to fix the problem. The Manufacturing department was responsible for processing the wafers within the standard cycle time. Manufacturing and production control were independent, and manufacturing did not control wafer starts.

Under this system, all went well if manufacturing kept up with all the wafers that were entering the fabs. Production control governed material starts independently of events in the fab. This inevitably inflated WIP inventories, and these inventories caused longer cycle times.

This approach's basic flaw is the assumption of a direct correlation between wafer starts and factory output. It assumes that starting more wafers will produce more output. This assumption is true in an unconstrained wafer fabrication line or other factory. Problems arise, however, when one pushes extra material into a manufacturing line that is operating close to its capacity. This simply generates WIP inventory. Excessive inventory in any manufacturing line causes many problems, including hidden quality problems and constant fire-fighting.

A New Manufacturing Philosophy

Eliyahu Goldratt's TOC thinking process (Goldratt and Cox 1992) reveals a basic core problem that still plagues many production environments. Formal and informal measurement systems focus on local performance at every work center. They do not distinguish between constrained and non-constrained operations.

TOC breaks through the local performance paradigm and creates a new manufacturing philosophy. This new approach accounts for manufacturing constraints and provides a systematic approach for solving production problems (see Table 9.2). This philosophy accords with the reality of the throughput world. SFM applies TOC to production, through

Table 9.2. Managing manufacturing constraints: Procedural steps (Goldratt and Cox 1992).

1. Identify the system's constraint(s).
2. Decide how to exploit the system constraint(s).
3. Subordinate everything else to the decisions made in step 2.
4. Elevate the system's constraint(s).
5. If the capacity of a constraint is elevated to the point that the constraint is broken, go back to step 1 and start the entire process over again.

the drum-buffer-rope (DBR) system. It is similar to just-in-time (JIT) manufacturing. Synchronous flow management is another term for JIT manufacturing.

Physical Constraints

Constraints limit the performance of all real systems. A constraint is anything that prevents a system from performing better or reaching its goal. Constraints can be physical, and a manufacturing station with insufficient capacity is an example of this. Lack of market demand is another physical constraint; the factory cannot sell something that no one wants. Problems with vendor deliveries also can constrain a process.

Goldratt and Cox (1992) use a Boy Scout hike to explain constraints. Each hiker represents a manufacturing operation. The troop's collective goal is to get from its starting point to the campground. It does not help if a few scouts arrive early; the goal is for everyone to arrive together. A scout named Herbie has a very heavy backpack, so he cannot walk as quickly as the others. Herbie is the constraint; the only way for the troop to go faster is to help Herbie go faster.

The Chinese general Sun Tzu (1983, 32) described constraints 2500 years ago. He described the drawback of a forced march: "The stronger men will be in front, the jaded ones will fall behind, and on this plan only one-tenth of your army will reach its destination." Here, the commander focuses on individual efficiencies and tells the soldiers to march as rapidly as possible. If an enemy is waiting, the enemy can defeat them individually as they arrive. The goal is to have the soldiers arrive as an army, not individuals.* Sun Tzu definitely knew about Goldratt and Cox's Herbie, or his equivalent.

In a factory, throughput occurs only when all operations have processed the work. In Goldratt and Cox's Boy Scout hike, throughput occurs only when all hikers have covered the ground. If the fastest hikers go as quickly as they can, the troop will string out along its path. The distance between the hikers represents WIP inventory (see Figure 9.2).

Managerial and Procedural Constraints

Procedural and managerial constraints are "do it (to) yourself" constraints. Managerial constraints include poor policy deployment, illogical measurement systems, and dysfunctional incentives. They also include preconceived attitudes about managing people, resources, or inventory. Behavioral constraints include work habits and attitudes. Examples include, "We've always done it that way" and "Keep busy or they'll lay me off." These constraints impede an organization from achieving its goal.

Identifying the Constraints

How does one find the constraint in a system? The factory must reduce inventory dramatically, since inventory can hide problems. Managers must observe the work area and process, and interact with employees. The organization can then identify the core problems. These might include performance measurements that encourage nonconstructive competition between shifts. Unmanageable batch sizes can starve the constraint while inflating inventory. Trying to optimize every process draws resources and organizational attention from their proper focus: the bottleneck.

---

*The commentator Tu Yü (Sun Tzu 1963, 99) described PERT-CPM (project evaluation review technique—critical path method) during the T'ang Dynasty (618–905 CE). "Now those skilled in war must know where and when a battle will be fought. They measure the roads and fix the date. They divide the army and march in separate columns. Those who are distant start first, those who are near by, later. Thus the meeting of troops from distances of [300 miles] takes place at the same time. It is like people coming to a city market."

That is, the planners estimate the time necessary to complete each phase of the project. The activities that will take the most time begin first. The separate project elements will, ideally, finish together to complete the project.

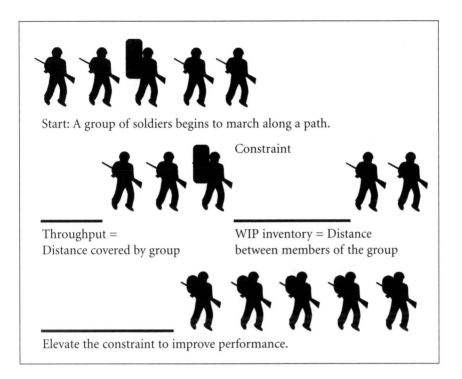

Start: A group of soldiers begins to march along a path.

Constraint

Throughput =
Distance covered by group

WIP inventory = Distance
between members of the group

Elevate the constraint to improve performance.

Figure 9.2. Throughput, constraints, and WIP in a hike or march.

Constraint Management

To manage a constrained process, a production manager must always think about opportunity costs. Opportunity costs result from forgone chances to sell a product and earn a profit. The cost accounting system does not treat opportunity costs; the company can't write them off as losses on its tax return. The managerial decision process must, however, always account for opportunity costs.

Marginal returns are another vital concept. What is the marginal, or differential, benefit of making and selling another piece? Think $\left(\dfrac{d\$}{dx}\right)_x$ where x is the current production level (see Figure 9.3).

Figure 9.3. Opportunity costs and constraint management.

The following production issues become important in a constrained process.

1. Scrap in or after the constraint is irrecoverable.

  —A bad workpiece's entry into the constraint wastes irreplaceable capacity.

  —Wafers with zero yields at final test have wasted the constraint's capacity. Chapter 7, on Mountaintop's zero scrap program, discusses this further.

2. Rework in the constraint is irrecoverable.

3. Idle time and partial loads in the constraint are irrecoverable wastes of capacity.

—Running out of work, or "starving the constraint," causes idle time.

—Unavailability of trained personnel causes idle time.

—Semiconductor wafers are "pallets" of several hundred or more electrical devices (die). A wafer with bad die is like a partially loaded pallet. When it enters the constraint, we are really running a partial load in the constraint. This is why wafer yields are very important in a constrained environment. Mountaintop's integrated yield management program, discussed in chapter 11, examines this further.

Constraints and Managerial Economics

Here is a simple example that shows how a constraint affects managerial economics. It uses a semiconductor wafer, but the example extends to other industries. The exception is the "partial load" analysis. Consider a traditional manufacturing process that uses pallets of a dozen pieces. Two pallets arrive at the constraint, and each has two defective pieces. The workers would discard the bad pieces, transfer two good pieces to a pallet, and process it. They'd wait until they could fill the other pallet with good pieces, thereby avoiding a partial load and wasted capacity. This option is not available in the semiconductor industry.

Consider a wafer that has 100 die (electrical devices) on it. It passes through 10 operations, of which the fifth is the constraint. The constraint processes four wafers an hour.

- The wafer (starting material) costs $50.

- Each operation uses $5 in materials and $5 in labor. The complete process uses $50 in material and $50 in labor, so the standard cost per wafer is $150.

- The die sell for $3 each, so a wafer with 100 good die earns $300 in revenue. The marginal profit, or throughput, on a good wafer is $300 less $100 for materials, or $200. (We are paying for the labor anyway. TOC treats labor as a fixed cost.)

Table 9.3 shows some transactions and their respective costs.

Table 9.3. Cost accounting and managerial economics in a constrained process.

	The wafer is scrapped after the fourth operation.		The wafer is scrapped after the sixth operation.	
	Cost accounting	Managerial economics	Cost accounting	Managerial economics
Material costs	$70	$70	$80	$80
Labor costs	$20	N/A; we're paying for it either way*	$30	N/A; we're paying for it either way
Marginal profit on a good wafer	Not applicable	N/A; we can replace the wafer	Not applicable	$200
Loss	$90	$70	$110	$280
Explanation (managerial economics)	We have wasted $70 in materials.		Instead of being ahead $200 (marginal profit), we are behind $80 (wasted materials).	

	The wafer is reworked during the sixth operation.		The wafer is reworked during the fifth operation (constraint).	
	Cost accounting	Managerial economics	Cost accounting	Managerial economics
Material costs	$5	$5	$5	$5
Labor costs	$5	N/A; we're paying for it either way	$5	N/A; we're paying for it either way
Marginal profit on a good wafer	Not applicable	N/A; we rework the wafer	Not applicable	$200†
Loss	$10	$5	$10	$205
Explanation (managerial economics)	This is as it appears. The marginal cost of the rework is $5.		Instead of being ahead $200 (marginal profit), we are behind $5 (marginal rework cost).	

*Unless the factory is paying for overtime. Then the extra labor becomes a variable cost, at $7.50 per operation (time and a half). Remember that we are looking at marginal costs and benefits.

†This is not a misprint. The rework uses a unit of the constraint's irreplaceable capacity. We permanently lose the opportunity to make a wafer and earn $200 marginal profit by selling it.

Table 9.3. *Continued.*

	Idle time: The workers take their scheduled hour for lunch together.		"Partial load": A wafer has 50 percent bad die at final test.	
	Cost accounting	Managerial economics	Cost accounting (treat half the wafer as scrap)	Managerial economics
Material costs	$0	$0	$50	$0
Labor costs	$0	N/A; we're paying for it either way	$25	N/A; we're paying for it either way
Marginal profit on a good wafer	Not applicable	4 @ $200	Not applicable	50 die @ $3.00 (marginal revenue)
Loss	$0	$800	$75	$150
Explanation (managerial economics)	Instead of earning $800 during the break, we have nothing. (Stagger lunch and break times to keep the constraint staffed and running.)		Instead of being ahead $200, we are ahead $50 ($150 in revenue minus $100 for materials).	

	The wafer is dead on arrival at final test with no yield.		The constraint runs out of work for two hours because a preceding operation stops.	
	Cost accounting	Managerial economics	Cost accounting	Managerial economics
Material costs	$100	$100	$0	
Labor costs	$50	N/A; we're paying for it either way	$0	N/A; we're paying for it either way
Marginal profit on a good wafer	Not applicable	$200	Not applicable	8 @ $200
Loss	$150	$300	$0	$1600
Explanation (managerial economics)	Instead of being ahead $200, we are behind $100.		This is similar to the workers taking a simultaneous break. Do not let the constraint run out of work!	

How to Handle the Constraint

Once the factory has identified the problems, it must decide how to control each major constraint's variation. Mountaintop used the DBR technique to control material starts. The plant also set up a buffer stock before a major system constraint to protect it against variation in the preceding operations. The preconstraint buffer is the exception to the plant's desire to get rid of WIP inventory. The constraint must never run out of work, since time losses at the constraint are irrecoverable.

To increase throughput, the factory must exploit the constraint. Employees must always be available to run the bottleneck. Aggressive cross-training allows the plant to have a series of backups to operate and maintain the constraint equipment. There should be inspections or tests immediately before the constraint to prevent bad pieces from entering the bottleneck. We must not waste the constraint's capacity on bad pieces.

Exploitation of the constraint yields two key lessons. First, there must be an adequate capacity cushion, or protective capacity, for all equipment and processes before the constraint. Reengineer batch processes for low inventory and rapid flow. Again, we want little or no inventory at any workstation except the constraint.

Second, we must focus on balancing product flow instead of line capacity. Smaller lot sizes require more setups and transfers. Operator or nonbottleneck equipment efficiencies have little relevance, because these do not affect overall throughput. The organization must look for the process and procedural constraints that impede product flow to the constraint, since faster flows allow lower buffer levels. Lower buffer levels reduce inventory carrying costs and allow quick responses to new customer orders.

"Push" Versus "Pull"

Mountaintop uses the DBR procedure to control production starts. DBR is similar to the "pull" system in JIT manufacturing. It supersedes the deficient "push" system in which the production control department pushes materials into the line.

Figure 9.4 illustrates the traditional production control system. The production control department releases materials into the line in response to sales forecasts or orders. This requires the manufacturing line to swallow large material batches. This system works for boa constrictors and pythons,

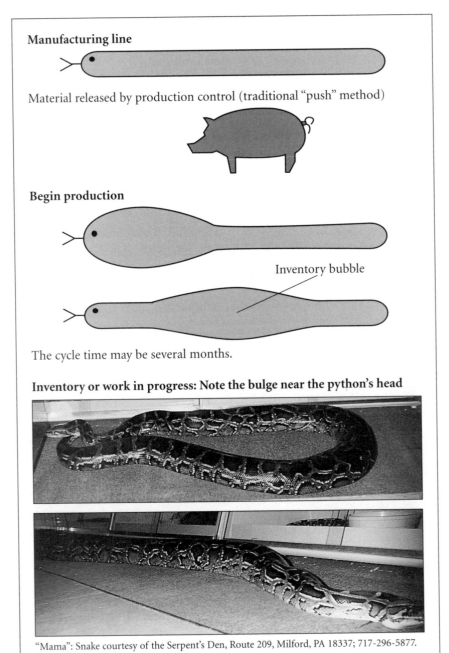

Manufacturing line

Material released by production control (traditional "push" method)

Begin production

Inventory bubble

The cycle time may be several months.

Inventory or work in progress: Note the bulge near the python's head

"Mama": Snake courtesy of the Serpent's Den, Route 209, Milford, PA 18337; 717-296-5877.

Figure 9.4. Traditional "push" manufacturing system.

whose digestive systems have evolved to process large meals. For others, it may result in serious indigestion and other disorders. Manufacturing managers may have to expedite high-priority jobs to alleviate the disorders. Also, boa constrictors may sleep for several months as they digest their large meals. The snake's digestive process has a long cycle time, which is something else we don't want in a manufacturing system.

DBR avoids this problem by ensuring a steady flow through the process. It leaves pig swallowing to systems that are designed to handle it.

DBR Manufacturing

Figure 9.5 demonstrates DBR with a hypothetical factory. This simple factory has only six machines, A through F, and it manufactures only one type of finished product. The figure shows the progression of work from raw materials to finished goods. Assume that there are no equipment breakdowns, rework, or scrap.

Suppose the factory followed the dogmatic strategy of maintaining high equipment efficiencies at all machines. Production control would release at least 20 units per hour of raw material to keep machine A busy. Machine B, however, cannot keep up with the flow from machine A. Machine C cannot keep up with B, and each station would accumulate five units of WIP inventory per hour. If this continues, the company will have to buy a warehouse to store the inventory. Goldratt and Cox (1992) describe this situation in *The Goal.* Not only does the company tie up money in inventory, it must pay to store it. Meanwhile, the hypothetical factory's management team was happy about the factory's "high efficiencies."

Since C can process only 10 units per hour, machines D, E, and F would accumulate no inventory. This factory cannot produce more than 10 units per hour, no matter how much production control pushes into

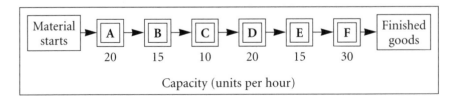

Figure 9.5. A hypothetical factory.

machine A. Machine C cannot process more than 10 per hour. If the market will absorb everything the factory can sell, machine C is the constraint. If the market will buy only eight units per hour, the market is the constraint. If the factory runs at full speed, the finished goods inventory will gain two units per hour.

Again, assume that the market will buy everything the factory can make. Under SFM, machine C "beats the drum"—it sets the pace at which the factory can "march." Machine C, not production control, decides how much material to start and when to start it. A *rope* links machine C to the material starts and limits these to what C can process. Under the JIT concept, machine C "pulls material into the line." In Figure 9.5, material release would occur at only 10 units per hour.

Idle Time: Good or Bad? Tying material starts to the constraint's output has far-reaching implications. If only 10 units start every hour, machine A's efficiency cannot exceed 50 percent. Metrics that call for high equipment efficiencies at all operations would cause confusion and make people resist the new method. For the change to succeed, then, everybody must realize that local efficiencies are not additive or even multiplicative. It is the constraint's efficiency that counts. If we improve the constraint's capacity until it is no longer the constraint, another operation assumes this role. If, in Figure 9.5, we find a way to double C's capacity, B and E become the constraints. Harris' Mountaintop plant accepts idle time at nonconstraints and calls it "opportunity time." Production associates can use this time for cross-training and improvement projects.

Idle time would be anathema in the cost world. It reduces the workers' and machines' efficiencies and raises the per-unit cost. The cost world wants the hourly workers to make product all the time and considers training an undesirable necessity. The idea of workers using time to reduce scrap (the zero scrap program) or interact with customers (customer contact teams) would appall the cost world. Chapters 6 and 7 on these Mountaintop programs show their immense benefits. Paradigms about per-unit costs and equipment efficiencies can prevent a company from realizing these benefits (see Figure 9.6).

The Constraint Is the Keystone. Intermittent idle time at nonconstraints does not affect the factory's throughput. *Idle time at the constraint, however, is irrecoverable.* Time losses at the constraint equal irremediable

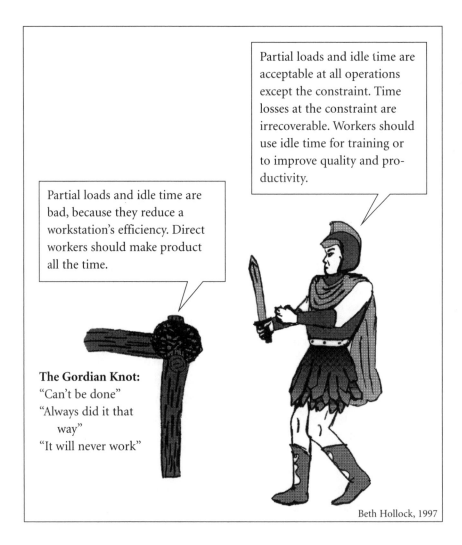

Figure 9.6. Partial loads and idle time.

throughput losses, so high efficiency is vital here. The constraint will be idle if it runs out of work, and trouble in preceding operations can cause this. Production rates may vary, equipment may break down, and there may be rework or scrap. The factory must never allow lack of work to idle the constraint. The new production solution would therefore be incomplete if it did not protect the constraint from starvation.

Buffers. Traditional factories keep inventory stocks at all operations to decouple them from the others. TOC tells us that only the constraint needs decoupling from preceding operations. To protect the constraint from starvation, the DBR system keeps a WIP inventory buffer in front of the constraint. The buffer decouples the constraint from the preceding operations and protects it from variation in those operations. In Figure 9.5, stoppage in machine B will not put machine C out of action. When B comes back on line, it can use its extra capacity to replenish the buffer. The buffer's size depends on the plant's experience with the processes before the constraint. If there is a lot of variation in those processes, the buffer must be large. Ask, "What is a realistic worst-case situation for the processes that come before the buffer?" Plan the buffer accordingly.

The buffer that protects the constraint from starving is the *internal constraint buffer*. A suitable internal constraint buffer should prevent the constraint from ever running out of work. However, variation after the constraint can make the factory miss shipment dates. To prevent this, factories also maintain *shipping buffers* of finished goods. Shipping buffers usually reside at the shipping dock, in case products do not arrive within the expected cycle time. If finished goods fall behind schedule, a hole will appear in the shipping buffer. This will prompt selective, proactive expediting of the parts in question to make sure the shipment date is met despite the problem.

Buffer management is the process of choosing appropriate sizes for the buffers and managing the overall production pipeline.

Buffer Management. After setup of the DBR system, we can learn a lot by observing and managing the internal constraint and shipping buffers. DBR calls for scheduling production only at the constraint, not at any other resource in the factory. DBR subordinates operations before the constraint to the constraint. Their mission is to keep work moving toward the constraint as quickly as possible. Operations after the constraint seek to turn WIP into finished goods as quickly as possible and get them to the shipping dock. Any disturbance in this flow is observable as variation in the internal constraint buffer or the shipping buffer. Stoppage in a pre-constraint operation will deplete the internal constraint buffer. Some of this variation simply reflects the inherent statistical fluctuations present in any production line. When should one intervene and take corrective action, if the buffer starts depleting?

The Goldratt Institute defines three levels of urgency for the constraint and shipping buffers. These are the OK zone, the watch-and-plan zone, and the act zone. The buffer should usually be two-thirds full. If it is always completely full, the factory has too much WIP inventory in the line.

Figure 9.7 is a schematic representation of a buffer. Inventory in the act zone is about to enter the constraint, or go to a customer. Inventory in the OK zone goes to the constraint, or customer, last. Work from the OK zone flows into the watch-and-plan zone, then into the act zone.

If work is missing from the OK zone, or if there is a hole in the OK zone of the buffers, there is no need for immediate intervention. This may simply be the result of normal production variation. A hole in the watch-and-plan zone means work that should have been at the constraint or shipping dock is missing from the buffer. Now, production personnel must plan corrective action in case the hole reaches the act zone. They must use the plan and intervene if the hole reaches the act zone, because this threatens the constraint's production or the shipping schedule. Timely intervention is often as simple as locating the work and rushing it through the remaining operations. This protects the schedules at the constraint and the shipping dock. This is careful and selective expediting, and it should happen infrequently. If such intervention happens frequently, the buffer is too small.

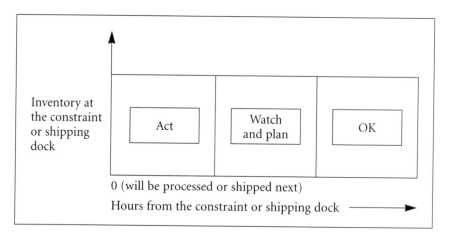

Figure 9.7. Buffer and status zones.

Buffer management is a very useful tool for production personnel, because it provides visibility to both the constraint and the shipping dock. There is always natural variation in a production pipeline. It is, however, a mistake to respond to every possible fluctuation. This will only result in massive confusion and continuous expediting. By learning how to manage the internal constraint and shipping buffers, shop floor personnel can deal with these random fluctuations much better. They can intervene proactively with the natural flow of WIP inventory only to protect the constraint and shipment schedules.

Employee Involvement and Education

Resistance to change is common in all organizations. Employees resist change because "we've always done it that way," and it is uncomfortable to do things differently. People are reluctant to admit they've been doing something wrong for 10 or 20 years. Drastic changes often require drastic paradigm shifts, and these are hard for people. When the change happens, there are no historical data to confirm the new approach. Managers, leaders, and change agents must work hard to guide the organization through radical changes. Every member of the organization must eventually buy into the change. It is the masses, ultimately, that make the change part of the organization's culture.

Change Management

Modern organizations are examining and trying a plethora of continuous improvement techniques. This can confuse and even frighten employees. Most employees do not get to participate in the analysis of new techniques that affect their work environments; a small group usually analyzes and assesses such techniques. Even if this group is very confident in the new approach, it must still convince the rest of the organization. The leaders at Harris Mountaintop adopted a two-pronged approach to persuade the entire workforce to accept SFM and make it successful. The approach relied on coherent communication and appropriate education for all employees.

Communication and Change

We cannot overemphasize the role of coherent communication in change management. Leaders must send consistent messages, because conflicting and ambiguous ones will confuse employees and reduce their confidence. If employees see evidence of doubt among their leaders, they will easily step back into the comfort zone of the old approach. Managing an organization is a lot like riding a horse. Horses are emotionally sensitive animals, and a rider who isn't confident can't make the horse believe otherwise. Rudyard Kipling (1982, 870) wrote, "They felt that the men on their backs were afraid of something. When horses once know *that,* all is over except the butchery." Leaders and change agents must therefore take every opportunity to show visible support for the new approach.

Senior managers at the Mountaintop facility met with all production and support personnel across the three shifts. They had open discussions on why the plant was adopting SFM. There was a consistent effort to involve employees with every step of the change. This allowed managers to identify employees' concerns, even if they couldn't resolve them immediately. All of the self-directed work teams had a chance to hear the same message directly from the most senior managers in the plant. A major advantage of SFM is that it is intuitively appealing, and it is common sense to most people. Enthusiasm soon began spreading across the entire plant.*

The Role of Training in Change

Like initial enthusiasm, knowledge is a key ingredient for making a radical change successful. Harris' Mountaintop plant therefore made significant investments in training and education. An external consultant gave four-hour overviews of SFM to all production personnel. These sessions gave

---

*This approach reinforces another lesson from our archetypal paradigm buster, Alexander the Great. There was a warhorse, Bucephalus, whom no one could ride. Ordinary Macedonian equestrians could not force the animal to obey them. Alexander, however, paid attention to the horse's concerns. The problem was that Bucephalus feared his own shadow. When Alexander turned him so he could no longer see it, he allowed Alexander to mount him. Bucephalus was then faithful to Alexander for the rest of his life. Change agents must similarly identify the organization's fears and help it surmount them. The organization will then follow willingly, where it would otherwise have balked.

employees their first significant exposure to synchronous flow concepts. Simple exercises and demonstrations helped employees challenge traditional assumptions about production. This helped employees put aside assumptions that were, to them, time-proven and even sacred. The plant started DBR in the wafer fabrication lines during these overview sessions.

There is another lesson here: People fear what they don't understand. Once they understand the change, they are willing to go forward.

These four-hour overviews gave employees enough knowledge to get the program going. The plant later gave comprehensive two-day workshops on production applications of TOC. These workshops provided greater detail and gave semiconductor industry-specific examples. These two-day workshops are from the Avraham Y. Goldratt Institute, which developed TOC. Four professional employees from Mountaintop completed the Goldratt Institute's training as "Jonahs." (Jonah is the consultant in Goldratt and Cox's (1992) *The Goal.*) An internal Jonah-Licensee conducted the two-day in-house workshops for all Harris Mountaintop employees.

Workshop participants use simulation software to manage a hypothetical factory. First, they try conventional production methods and techniques. These focus on every production station as an isolated resource. The students quickly discover that they cannot make the desired production schedule this way. The simulated factories also fill with WIP inventory very quickly.

The workshop next applies the DBR solution to the same factory. The instructor introduces buffer management, and simulations help the students understand it. Workshop participants begin with their existing set of assumptions about manufacturing. They emerge with a basic understanding of SFM and DBR. Every Mountaintop employee from production, design, process and product engineering, accounting, human resources, and facilities support attended this workshop. This massive investment in education resulted in a much better universal understanding of SFM. It also reinforced Harris Semiconductor's commitment to the SFM philosophy. With this basic understanding, people have found it much easier to overcome their initial resistance to change and adopt the new production philosophy.

The orientation program for all new hires at Harris Mountaintop now includes an overview of TOC and SFM. Senior management continues to highlight the improvements in throughput and cycle time that have come about since the introduction of SFM. The terms *drum, buffer,* and *rope* are now part of the normal vocabulary for all Harris Mountaintop employees.

Challenging Traditional Assumptions in Production

During the transition to SFM, Mountaintop had to reexamine and challenge many assumptions that have governed production. Conventional wisdom has changed in the global manufacturing arena. Here is a brief description of some key assumptions that have changed radically.

Inventory

Conventional wisdom calls for enough inventory to keep every work center busy all the time. Inventory isolates individual work center variations, or decouples the workstations. If a machine stops working, there is enough WIP inventory to keep the ones downstream going. This approach relies on the assumption that the factory can sell all of the inventory. This assumption relies, in turn, on the accuracy of long-term sales forecasts.

JIT manufacturing highlighted the importance of short cycle times. Excessive inventory leads to longer queues in front of work centers and increases the overall production cycle time. If the production cycle time is too long, marketing and sales have to rely more heavily on forecasts to predict the needs of customers. Long-term forecasts are rarely accurate, as most production managers know very well. Factories with high inventories end up producing goods that no one wants while missing shipment dates for products that customers do want. High inventories also hide quality problems, which often leads to major rework or scrap losses in the production line.

SFM emphasizes the need for flow in production lines. Smooth flow in the production pipeline is possible only if the factory controls WIP. Meanwhile, throughput, by definition, includes only finished goods that customers want. DBR manufacturing is a systematic way to limit inventory to what is necessary to achieve continuous flow. Small inventory levels permit faster production cycle times and better shipment performance. Conventional wisdom has changed: Low inventory is better, and it promotes customer satisfaction.

Batching and Setups

Traditional manufacturing systems always tried to assemble large batches of similar products. The workstation could then process them after a single setup. Production managers focused on long runs and as few setups as possible. The idea was to keep the machines working, and setups are downtime that reduces utilization and efficiency. The abundant WIP queues at each station made this easy. Remember, however, that the organization's goal is not to achieve high equipment efficiencies. It is to make money now and in the future. Assumptions about efficiencies and utilizations are self-limiting paradigms that keep the organization from achieving this goal. Instead, we must ask, "How do fewer setups and longer runs at nonconstraints improve throughput?" The answer is, they don't. They promote local efficiencies and optimization, which have no effect on the overall throughput.

In the low inventory environment of the throughput world, SFM repeatedly emphasizes the disadvantages of batching products. If the capacity-constrained work center had to wait for large batches to avoid setups, the idle time would cause huge revenue losses. If stations before the constraint wait for full loads, the constraint might run out of work. We have already seen that low efficiencies at nonconstraints are not a problem. Stoppage at the constraint is, however, a disaster.

After the constraint, workstations must process the work as rapidly as possible to get it out the door. If they chase the illusory goal of "fewer setups," the cycle time will be longer. It is better to accept frequent setups and deliver the product on time. If the shipments are on time, customers will not complain about "too many setups." If shipments are late, customers will not care if the plant had long, "cost-efficient" production runs. Conventional wisdom has changed. Large batches are bad for flow, and fewer setups at nonconstraints do not help the organization become more profitable (see Figure 9.8).

Small Batches and Customers

Conventional wisdom calls for deliveries of large batches, or complete orders, to customers. It is more efficient, for example, to ship one full truckload than five partial ones. JIT philosophy, however, stands conventional wisdom on its head.

Figure 9.8. The long production run/low setup paradigm.

Tu Mu (803–852 C.E.), a commentator on Sun Tzu's *The Art of War* (1963, 103–104), introduced the idea of partial deliveries and continuous flow.

> *You select the most robust man of ten to go first while you order the remainder to follow in the rear. So of ten thousand men you select one thousand who will arrive at dawn. The remainder will arrive continuously, some in late morning*

and some in mid-afternoon, so that none is exhausted and all arrive in succession to join those who preceded them. The sound of their marching is uninterrupted. In contending for advantage, it must be for a strategically critical point. Then, even one thousand will be sufficient to defend it until those who follow arrive.*

Here, the idea is to hold a critical position until the entire army can arrive. Now suppose we are talking about work pieces instead of soldiers, and the customer needs 10,000. The first 1000 will keep the customer's factory busy until the next 1000 can arrive, and so on. If we try to deliver the full 10,000, the shipment may arrive late. The customer will have to wait for the delivery before it can do any work. When the 10,000 parts get there, the customer will have to store them. Many customers are, in fact, asking for partial shipments under the JIT philosophy.

Readers of Goldratt and Cox's *The Goal* (1992, 245–246) will immediately recognize this scenario. It is the strategy that UniCo used to get a contract for 1000 Model 12s for delivery to Bucky Burnside. Alex Rogo, UniCo's plant manager, says, "There is no way we can deliver the full 1000 units in two weeks. But we can ship 250 per week to them for four weeks." The sales manager then discovers, "They like the smaller shipments even *better* than getting all 1000 units at once!"

Equipment and Labor Efficiencies

We have seen how cost accounting systems allocate labor and overhead to product costs. High efficiencies spread the fixed costs over more products and reduce the cost per piece. An implicit assumption behind this argument is that higher labor and equipment efficiencies at all work centers translate to more finished goods. These must be salable, however, to improve the factory's revenues. These assumptions have led cost accounting systems to measure efficiencies and variances for individual operations. SFM invalidates these assumptions (see Figure 9.1). High equipment and labor efficiencies at nonconstraints do not translate into more throughput.

---

*In the preceding hike/march example, the distance between hikers is inventory. When all the hikers arrive together, there is no WIP. Here, the soldiers represent products and not operations, and their intermittent arrivals are partial deliveries.

Balanced Capacities and Other Mythical Entities

Traditional wisdom calls for balancing capacities across all workstations. Factories rarely achieve this in practice, and those that do always struggle with shifting bottlenecks. This is because any fluctuation in the manufacturing pipeline produces a temporary constraint in product flow. Goldratt and Cox (1992, 104–112) show what happens under these conditions. The example uses a series of workstations that can process an average of 3.5 pieces per turn. Each workstation ships the quantity shown by a die roll (one to six). It cannot, however, process more work than it has waiting. The example shows that the "balanced" production line generates fewer than 3.5 pieces per turn.

If the market will absorb everything that the plant can make, there will usually be a single physical constraint in the process. Efforts must focus on exploiting and elevating this constraint. Trying for high efficiencies at nonconstraints is, at best, a waste of time. We have seen that it will actually hurt the factory's performance by generating piles of WIP inventory. Dysfunctional performance measurements that focus on local efficiencies will produce exactly that: suboptimization. This will tie up cash in inventory, and probably hurt cycle time and delivery schedules. Remember the warning about performance measurements: "Be careful what you wish for; you might (and probably will) get it."

Conventional wisdom has changed. Seeking high labor and equipment efficiencies at all work centers counteracts the organization's goal of making money by delivering throughput. Efficiency metrics must focus on the constraint, because the constraint governs throughput. We cannot overemphasize the message, "Time losses at the constraint are irrecoverable." Nonconstraints must subordinate their efforts to the welfare of the constraint. This means providing a steady flow of work to the constraint and to the shipping dock.

All wafer fabrication facilities at Harris Semiconductor Mountaintop, follow the pipeline approach. One person is responsible for managing the entire set of production operations within a single fab. This avoids having managers for different operations compete for the highest local efficiency.

Continuous Improvement

Harris Semiconductor's Mountaintop plant subscribes to the philosophy of continuous improvement. *Kaizen,* or continuous improvement, is a requirement for any journey to excellence. A key feature of Mountaintop's continuous improvement, however, is the coordination of the key improvement tools. Quality improvement programs and activities do not work in isolation. Instead, techniques like SFM, total productive maintenance (TPM), and integrated yield management (IYM) support and reinforce each other (see Figure 9.9).

To reap the greatest benefits from SFM, we must focus improvement efforts appropriately. These efforts can address equipment and process issues. TPM involves small production teams in improving the effectiveness of production equipment. IYM focuses on the effectiveness of semiconductor fabrication processes. IYM's goal is to improve probe test yields at the end of wafer processing. Details of TPM and IYM appear in chapters 8 and 11 in this text. SFM is a separate tool that provides the underlying philosophy that governs the production line. All three, independently, can improve the overall performance of the manufacturing system somewhat. The challenge at Harris Mountaintop was to synergize these three tools so that their result would exceed the sum of their parts.

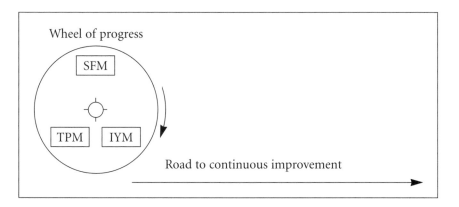

Figure 9.9. Relationship between continuous improvement tools.

Program Synergies

TPM can improve the constraint's overall equipment effectiveness (OEE). This translates into more throughput and profit. Yield improvement translates into pure profit, because the plant has already spent the money to make the wafers. A wafer with a 100 percent yield has the same variable cost as one with 80 percent yield. When the factory, not the market, is the constraint, the extra 20 percent is pure profit (see Table 9.4). IYM synergizes with SFM, because SFM tells us that we cannot simply make more wafers to satisfy the demand. We must get as much yield as possible from each wafer. The zero scrap program similarly supports SFM, and Table 9.3 shows why it is important.

Table 9.4 relates IYM and TOC. Again, suppose that a wafer costs $100 in materials and $50 in labor. A wafer has 100 die, each of which sell for $3.00. The factory can make 100 wafers a day.

Inherent conflicts between improvement philosophies often cause confusion for shop floor personnel. These people are ultimately responsible for performing the changes and improvements. Supporters of one approach resist the introduction of another, since they don't want to reinvent the wheel. The leaders at Mountaintop therefore spent time to teach the workforce the interactions between TPM, IYM, and SFM. They recognized the intrinsic advantages of each improvement tool and showed how each tool supports the organization's goals.

Total Productive Maintenance and SFM

The goal of business organizations is to make money now and in the future. SFM recognizes this goal and the emergence of the throughput world. Through the DBR approach, SFM provides the overall production philosophy for the manufacturing system. For DBR to be effective, key pieces of equipment—especially the constraint—need to perform consistently. TPM provides a well-documented set of techniques and procedures to get effective equipment performance through small group activities. In isolation, TPM would try to reduce downtime and breakdowns at every operation. This approach can achieve small improvements everywhere, but excellence nowhere.

Table 9.4. IYM and the Theory of Constraints.

	Case 1: The market will buy up to 8000 die. (The market is the constraint.)			
	Make 80 wafers with 100% yield (8000 die).		Make 100 wafers with 80% yield (8000 die).	
	Cost accounting (CA)	Managerial economics (ME)	Cost accounting (CA)	Managerial economics (ME)
Material costs	$8000	$8000	$10,000	$10,000
Labor costs	$4000	N/A; we're paying for it either way	$5000	N/A; we're paying for it either way
Sales	$24,000	$24,000	$24,000	$24,000
Gain (CA) or marginal gain (ME)	$12,000	$16,000	$9,000	$14,000
Explanation	In both the cost world and throughput world, the penalty for lower yield is the cost of making more wafers to fill the demand. In the cost world, this is $3000 for labor and materials. Under the managerial economics model, the variable cost of 20 extra wafers is $2000.			

	Case 2: The market will buy up to 12,000 die. (The factory is the constraint.)			
	Make 100 wafers with 100% yield (10,000 die).		Make 100 wafers with 80% yield (8000 die).	
	Cost accounting (CA)	Managerial economics (ME)	Cost accounting (CA)	Managerial economics (ME)
Material costs	$10,000	$10,000	$10,000	$10,000
Labor costs	$5000	N/A; we're paying for it either way	$5000	N/A; we're paying for it either way
Sales	$30,000	$30,000	$24,000	$24,000
Gain (CA) or marginal gain (ME)	$15,000	$20,000	$9,000	$14,000
Explanation	In both the cost world and throughput world, the 20% lower yield costs $6000, or *the market price of 2000 die.* Compare to $3000 (cost accounting) and $2000 (managerial economics) when the market was the constraint. When the factory is the constraint, the lower yield costs an irretrievable opportunity to sell 2000 die.			

TPM: Focus Organizational Resources on the Decisive Operation

SFM applies the guidance of Frederick the Great: "One who tries to protect everything protects nothing." SFM focuses TPM on the constraints and near-constraints in the production line. We recognize that we do not have unlimited resources, such as technicians and equipment repairers. We must concentrate these resources at the decisive part of the process: the constraint. We accept downtime at nonconstraints if that is the price for protecting the constraint. We will accept lower availabilities, rate efficiencies, and operating efficiencies at nonconstraints. (Imperfect quality rates after the constraint are not acceptable, since those losses are irreplaceable.) The message is, "Protect the constraint's efficiency at (almost) any cost." Selective TPM activities, instead of across-the-board efforts, can produce large benefits very quickly.

Equipment performance has a natural link to the effectiveness of the process. When IYM investigations reveal consistent yield losses from a particular process, TPM improvement efforts can immediately focus on that operation. SFM tells us that yield losses after the constraint are particularly bad. They represent not only scrap, but irreplaceable capacity and throughput. IYM can therefore guide TPM efforts according to the broad operating perspective that comes from SFM.

These three improvement tools—SFM, IYM, and TPM—are all part of the wheel of progress that carries organizations along the road to continuous improvement. When they work together, they give a consistent message to shop floor personnel. They encourage every member of the manufacturing organization to consider the global picture instead of local optimization. This, by itself, is a breakthrough in organizational thinking.

The Market-Constrained Environment

So far, we've examined the very desirable situation in which we can sell everything we can make. When a factory can make more finished goods than it can sell, the market is the constraint. Lack of orders, not equipment capacity, now limits the factory's throughput. The market, and not a manufacturing operation, now beats the drum that sets the factory's pace.

Manufacturers are familiar with cyclic demand and business cycles. The overall world economy has been growing consistently. Peaks and valleys in this growth, however, have repeatedly shifted manufacturing system constraints between the factory and the market. DBR is a powerful tool for governing production when there is a physical constraint in the factory. What happens, though, when the constraint shifts to the market?

Dysfunctional Reactions to Market Constraints

When the market turns soft, traditional manufacturing facilities try to keep busy. To do this, they rely heavily on long-term demand forecasts. Forecasts, by definition, are inaccurate, and long-term ones are the worst. Everyone knows this, but the cost world forces many production managers to react to "inadequately loaded" factories, poor equipment efficiencies, and "unabsorbed costs." Under market-constrained conditions, the factory should concentrate on keeping existing customers happy. It can do this with short production lead times and outstanding due-date delivery performance. Instead, traditional factories generate excessive WIP inventory in their pursuit of local efficiencies and cost absorption. The factory ceases to distinguish between WIP for customer orders and WIP for an unreliable forecast. The forecast becomes a deadly Siren whose song lures the company onto the shoals of bloated inventories, late shipments and unhappy customers, and continuous expediting. Odysseus, or Ulysses, had the right idea when he met the Sirens. He ordered his sailors to plug their ears with wax so that no one could hear them (see Figure 9.10).

Traditional manufacturing facilities have trouble breaking out of this vicious cycle that recurs whenever the constraint shifts into the market. SFM and TOC, however, offer an entirely different perspective.

Operation in a Market-Constrained Environment

When the factory determines that the market is the constraint, it must reexamine the manufacturing system approach. It must overcome inertia in the system quickly. The first manufacturing organization that adapts to its new environment should beat its competitors to the goal. Synchronous flow of WIP inventory is still of prime importance in the production facility. At Mountaintop, the DBR system provides many advantages even in a market-constrained environment. While the details are a little different, the operating philosophy of the system is basically the same.

To exploit the market constraint, the entire manufacturing system must become even more responsive to its customers. Quality has become a

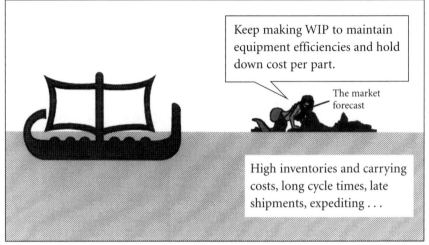

Keep making WIP to maintain equipment efficiencies and hold down cost per part.

The market forecast

High inventories and carrying costs, long cycle times, late shipments, expediting . . .

Mermaid/Siren adapted from Totem Graphics clip art

Figure 9.10. Dysfunctional goals in a market-constrained environment.

prerequisite for competing in global markets. Delivery and customer service also are critical competitive factors. Lead times that are shorter than competitors', and exceptional on-time delivery performance, should become the factory's top priorities. Although the market is the constraint, the manufacturing system can still use the DBR approach as its operating philosophy. The former physical constraint now becomes the control point in the production pipeline. The factory can still synchronize material releases with the control point's output. Since all production stations now have extra capacity, the factory can afford to shrink the internal constraint and shipping buffers. This will further reduce production cycle times.

The factory should start only those raw materials that are necessary to meet actual customer demand. It must not tie up its manufacturing capacity by producing unsalable inventory. In a market-constrained situation, short production lead times become a distinct competitive advantage. Chasing long-term forecasts with useless inventory lengthens these lead times and undermines this advantage; focus instead on actions that will promote short-term sales. SFM helps factory personnel sort through the jungle of productivity metrics and focus on the most important one: throughput. Throughput, or actual sales to customers, is what counts.

Project Management

The Theory of Constraints also applies to project management, and Mountaintop recently completed a new semiconductor factory in record time. To do this, Mountaintop had to shatter a paradigm about factory startups (see Figure 9.11). This section will discuss the application of TOC to project management.

Figure 9.11. Factory startups: Cost and throughput worlds.

General Carl von Clausewitz (1976) wrote that business is war: a competition between organizations. The famous former coach of the Green Bay Packers, Vince Lombardi, said that business, football, and war are similar. The idea is to win, to defeat the opponent. Unlike war and football, business is not always a win-lose activity. Instead of taking someone else's pie, a business can create a new one. However, the same principle applies when contending for an existing market or creating a new one: Get there first.

Sun Tzu (1963, 1983) repeatedly stressed the importance of speed. He wrote, "Speed is the essence of war" and "When campaigning, be swift as the wind." The first army to arrive on the battlefield has an advantage over those who arrive later. The first company to get to the marketplace also has an advantage. When its competitors arrive, it has already established a reputation and customer relationships. The new arrivals must then fight uphill to get a share of the market. When the game is King of the Hill, it helps to start at the top.

By breaking self-limiting paradigms, we can do this and achieve seemingly impossible results. The Carthaginian general Hannibal led an army, including war elephants, across the Alps. Difficulties with supplies, and hostile terrain and weather, made the Alps an "obviously" impassable barrier. Hannibal's reaction was, "We will either find a way, or make one." No one knows the Roman commander's last words at the battle of Cannae, but they are easy to guess: "Where did all the Carthaginians come from?" In Mountaintop's case, the idea was to build a factory to process eight-inch (200 mm) silicon wafers. The expansion project's name was "Project Raptor," after the velociraptor ("swift predator") dinosaur. Other companies also are building such factories, but ours began making wafers in January 1997. Figure 9.12 shows what happened.

The cost model would reject the parallel activities under Project Raptor, and a classroom accounting exercise would agree. The project incurs higher initial costs, because it hires workers and buys equipment before the factory is complete. The goal, however, is to make money by making products. The parallel activities let us start doing this much sooner. When the competitors' "less expensive" factories finally come online, ours will have been selling products for several months. Organizations that

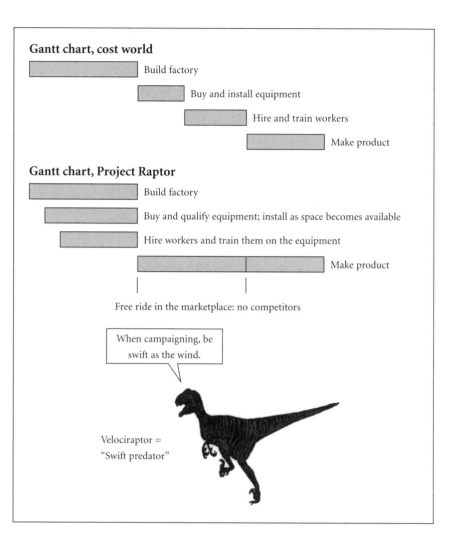

Figure 9.12. Factory startups: Cost and throughput worlds.

work in the cost world do not recognize the enormous opportunity costs that result from delays in factory completion.

Eliyahu Goldratt's new book, *Critical Chain* (1997), applies the Theory of Constraints to project planning.

CHAPTER 10

ISO 9000 and QS-9000

Clinton Chamberlin, John Benjamin, and
Robert F. Longenberger

What are ISO 9000 and QS-9000, why are they important, and how does Harris Mountaintop approach them? Some companies view ISO 9000 as a costly, time-consuming nuisance. They want to get the registration certificate so they can do business in Europe. Similarly, some companies treat QS-9000 as a necessary evil; they have to comply with it to sell parts to the automotive industry.

Harris' Mountaintop plant, and other progressive factories, recognize ISO 9000 as a systematic program for assessment and improvement of a quality system. It does not matter whether our customers require ISO or QS-9000 registration. The programs' discipline helps us identify and correct weaknesses in our quality system and improve it. ISO 9000 and QS-9000 are valuable investments that repay the time and effort spent on them many times over (see Figure 10.1). Their systematic assessment of the quality management system can improve productivity and quality and help the company make money (Scotto 1996).

Road Map to ISO 9000 Certification

The Decision

In late 1992, the division management of Harris Semiconductor decided that certification to the ISO 9000 standard would be important to the division's future success. In Mountaintop, questions arose. Who or what is ISO? Don't we have enough to do already, without more audits? What's

The ISO 9000 series standards and QS-9000 requirements are useful, systematic frameworks for assessing a quality system. We can use them to improve our quality, productivity, and profitability.

ISO 9000 registration is something we have to do if we want to do business in Europe. QS-9000 is something we have to do if we want to sell to the automotive industry. Both programs are costly and time-consuming annoyances.

The Gordian Knot:
"Can't be done"
"Always did it that way"
"It will never work"

Beth Hollock, 1997

Figure 10.1. Misconceptions about ISO 9000 and QS-9000.

wrong with our existing quality system? Will it really benefit our business? The road to achieving certification would be difficult, but looking both forward and back, the benefits are clear.

ISO is the International Organization for Standardization, which is located in Geneva, Switzerland. The ISO 9000 standards for quality are only one of the series of standards this organization generates. ISO is not a

European organization; most countries of the world, including the United States, belong to it through the representation of ISO member bodies.

The initial concern was that the new European Community would use the ISO 9000 standards to limit imports from non-European countries. The alternative, positive view was that the standards would serve as a common denominator for quality systems. Standardization would encourage international trade and provide internal benefits to an organization.

Mountaintop saw three distinct benefits of the program.

1. External benefits

   —Retention of existing customers

   —Gaining new customers, especially in Europe

   —Entry into new markets

   —Fewer dissatisfied customers

2. Internal benefits

   —Greater control of business

   —Better internal discipline

3. Quality cost benefits

   —Reduce scrap and waste

Picking a Registrar

The selection of the registrar was made at the Harris Semiconductor division level with inputs from the operating units. The choice involved three criteria.

1. Experience within our industry. The registrar must understand the issues "appropriate to the business" that arise in the interpretation of the standards.

2. Reputation within the quality community. Other organizations must view our certification as well-earned and not a rubber stamp.

3. The number of countries in which the registrar has accreditation.

There is no central ISO organization that grants certification. The certification process is a three-tier system. An accrediting body in each country oversees third-party organizations known as registrars. In Europe, many of these accreditation groups are either governmental or quasi-governmental such as the Dutch Council for Certification (RvC). The primary organizations in the United States are the Registrar Accreditation Board (RAB), which is an American Society for Quality affiliate, and the American National Standards Institute (ANSI).

The registrars themselves are independent certifying bodies that do not do business with the companies they audit. Independence and impartiality are key aspects of a valid audit. Customers can audit their suppliers; these are second-party audits. Registrars, who do not buy anything from their auditees, perform third-party audits.

The final tier is the auditors who work for the certifying bodies. This complex structure is important to understand. Accrediting bodies in some countries do not always recognize registrars with accreditation in others.

Harris chose Det Norske Veritas (DNV). DNV has experience in our industry, a good reputation in the quality community, and accreditation in seven countries with two pending. The division looked at cost, but did not worry about minor cost differences between prospective registrars. Harris' internal efforts would far outweigh the registration costs. The division also considered the registrar's pool of well-trained, experienced auditors. This factor is especially important as the number of ISO and QS certifications rises. Good auditors are at a premium.

QS-9000's Influence on Registrar Selection

QS-9000 is a new, and very important, consideration. As the North American automakers begin to require third-party audits for their joint QS-9000 requirement, the registrar's strength becomes even more important. DNV was an organization of sufficient size and structure that chose to offer QS-9000 certifications. It had the resources to train auditors and receive recognition from the automotive industry. The section on QS-9000 further discusses the automotive quality requirements.

Team Approach

While registrar selection was in progress, total quality management (TQM) was spreading through every level of the Mountaintop organization. Several years before this period, the quality department would have driven ISO 9000 certification. Plant management, however, decided that this would be part of the "Big Q" approach, not just "little q." ISO compliance efforts would focus the plant on working as a group to better meet the needs of customers (see Figure 10.2). It would not be a mere exercise in passing a quality criterion to get a certificate.

ISO 9000 covers every aspect of the company's operations. Diligent compliance helps the company satisfy its customers. While a manager or department can lead the company through the process, every department must actively participate.

ISO 9000 registration is a quality management program, and it affects the company's quality system. It is the quality department's job to achieve and maintain certification.

The Gordian Knot:
"Can't be done"
"Always did it that way"
"It will never work"

Beth Hollock, 1997

Figure 10.2. ISO 9000's role in the organization.

This organizationwide effort began with the selection of a steering committee to plan, direct, and monitor the certification effort. It had representatives from all areas of the factory: manufacturing, engineering, purchasing, quality, and human resources. The original leader was the manager of human resources, not the quality manager. This was, in retrospect, an excellent decision. *Achieving certification is much less about reaching quality goals than it is about having the whole organization own the process.* Yes, the quality organization could lead the drive for certification, and succeed. Without the whole organization's active involvement, however, it is harder and it produces fewer long-term benefits. We want ISO 9000 to become part of the organizational culture, and the best way to do this is to involve everybody.

About six months before submission of the quality manual for review, the ISO steering committee selected point people for each system element. The point people would be in charge of writing each section of the manual. For simplicity, each section of the quality manual would reflect an ISO system element. The point people searched the organization for the people they felt could best form a team to write their sections. They received copies of the ISO standard, copies of their respective sections from Harris and external companies' quality manuals, and other interpretation information. They would be responsible for the new Mountaintop quality system. It would reflect both the ISO 9000 requirements and the way the team wanted and expected to do business. The team would rewrite the quality manual from start to finish. The previous manual had grown like the tax code for nearly 35 years. It was time to simplify it and reflect current best practice. This also required review and writing of second-tier (see Figure 10.3) or general plantwide instructions.

The third-tier documents, or specific work instructions, needed review and updates. The best people to do this were the people who actually used them. Did these documents reflect current best practice? This is what we expected from our new quality manual, and we expected it from the work instructions, too. The plant steering committee made it a goal for operating teams to review and update their own operating instructions. Final review was the responsibility of the technical community. Placement of the actual activity in the hands of our production associates, however, raised their ownership of the quality system. Again, we want to involve

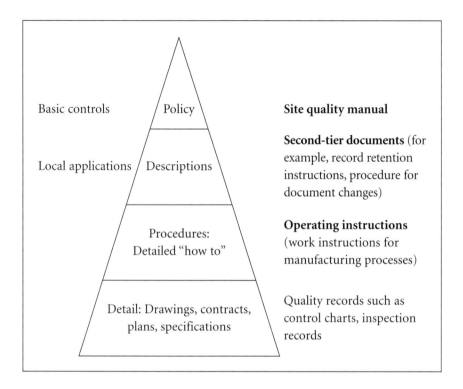

Figure 10.3. Document hierarchy: Quality system triangle (Arter 1994, 19).

everybody in ISO 9000. When people participate, they take ownership. This would not have happened if engineering had kept complete owner-ship of the documentation process.

Establishing Timing and Goals

To set the tone and communicate the goals for this project, the plant man-ager met with all the section team leaders and their teams. The message was clear.

1. The certification process was essential to the future of the plant and the company. The Semiconductor Sector's management team decided that all domestic and foreign plants must receive certifica-tion to compete in the global markets we serve.

2. The process must not reduce our ability to serve our customers; we needed to maintain a high on-time-delivery rating while we were achieving certification.

3. Keep it simple.

4. The goal was certification in 12 to 16 months.

The plant manager also held an all-employee communications meeting to make sure that everyone understood the importance and management commitment.

The selection of the ISO management representative was also critical. The paradigm had to change: ISO 9000 is not merely another quality system that the quality department manages (see Figure 10.2). This led to the selection of an individual from the process engineering group. He had the authority to make changes, ensure the proper corrective actions, and provide a single interface to the registrar for all issues.

To reinforce the paradigm shift, the manager of the human relations department led the ISO steering committee. Managers from most organizations were selected for the team. This committee members would select the leaders for each of the ISO standard sections and would set the timing and goals for the milestones. They also spent a lot of time selecting the individual members of the internal audit team. In the beginning, the committee met twice a week and reviewed the team progress monthly. Later, meetings became biweekly as the elements fell into place.

The Internal Audit

Before assessment for certification, the organization must perform an internal audit. Selection of internal auditors is very important. These people must audit in areas with which they are unfamiliar. They must understand the relationship between the quality system and the ISO standards. Finally, and most importantly, they must work at and with all levels of the organization.

The ISO steering committee selected eight people as the first internal auditors. They would perform this role in addition to their normal job functions. They included engineers, technicians, and a production associate. Only one had previous auditing experience. Since theirs was probably the most difficult task, a member of the ISO steering committee was assigned to help them. Lack of full-time auditors places a strain on the part-time auditors. However, part-time auditors provide a wide cross-section of skills and experience. Auditees view them as part of an effort to improve the entire organization. The auditors are not people from the quality department who look for violations.

The auditors received two days of training, as did the ISO steering committee. The training covered ISO, interpretation of the standard, and how to be a good auditor. Harris' corporate quality department presented the course. It was especially useful as a jump-start to the process.

Development of internal audit methodology followed the training. The plant was broken down into 27 distinct areas. For each of these areas, a review was held to determine which sections of the Mountaintop quality system applied in each area.

Guideline questions for each system element were also developed to provide an easy reference for the auditors during their audits. The auditors worked in teams of two, which allowed discussion and second opinions on possible internal nonconformances. Experienced auditors can work alone, but this is not a good practice for inexperienced auditors. Finally, the management representative for ISO 9000 was responsible for the internal audit activity. This is very appropriate, because the internal audits are key in determining if the quality system is effective.

The first cycle, including corrective action, preceded the pre-assessment. A very helpful aspect of the pre-assessment was the training opportunity it provided the internal audit team. Working with an experienced ISO auditor taught the fine difference between determining whether a system or procedure existed and the more subtle aspects of its effective use.

After the successful certification process, Mountaintop made several minor changes in the internal audit activity. The internal audit teams now have at least three people on them. This puts less burden on any one individual. An audit of any of the 27 areas must have at least two team members present. However, the added flexibility promotes timely auditing. Also, new auditors are added every year, and those who have had the extra responsibility for several years can retire. New auditors go through the two-day internal course and then join teams with auditors who have one or two years of experience. Finally, new employees with about one year of experience are receiving assignments as auditors. They have their feet wet in Mountaintop, but often have not ventured far from their immediate job function. They must audit and understand areas of the plant from which they are independent. This lets them see other aspects of what helps us succeed. They must also understand our quality system very well as an auditor, which gives them the "Big Q" view of the plant. They learn all the factors involved in meeting our customers' needs and driving our business to further success.

Keep It Simple

The manufacturing process at Mountaintop is highly technical, and a large group of experienced engineers define it. Engineers often create instructions with a lot of detail and structure. This is a recipe for failure in the ISO world of "Say what you do, and do what you say." In the words of the plant manager, "If you say that you will always get up on the left side of the bed and turn right to go to the kitchen for breakfast, you will surely fail to follow that operating instruction more than once." Instead, say, "Breakfast material is in the kitchen, but you need not have breakfast immediately upon awakening." This defines where to go to perform the job correctly, but it allows some room for deviations. Deviations are not acceptable if they can affect product quality, but unimportant technicalities must not turn into ISO nonconformances.

It is difficult to think this way. The ISO steering committee had to review the new quality manual and all of the second-tier documents several times. The committee could not do this with all the individual operating instructions. It delegated responsibility for rewriting the instructions to the operators who perform the jobs. This works well if the operators are experienced and can work as a local area team. They can remove unnecessary detail and clutter from the instructions while preserving their essence. The plant got rid of more than 30 percent of the existing instructions. The responsible engineer reviewed the new instructions and made any necessary corrections before approving them.

The Pre-Assessment

The pre-assessment is optional, but it is among the most important elements of successful implementation. For a fee, the registrar will visit the site and examine every aspect of your compliance to the standard. The benefits include the following opportunities.

1. Learn how the standard's interpretation applies to *your* business. The requirements are intentionally general, to cover any type of business activity. The registrar will provide invaluable insight in this area during the pre-assessment.

2. Learn how the process works.

3. Learn the expectations of your registrar, and what the "hot buttons" are.

4. Find the deficiencies that you overlooked.

5. Allow the internal audit team to learn how to apply the standard and what questions to ask.

6. Get some advice.

The last point requires explanation. During the pre-assessment, you are paying the registrar's representatives to help you understand the standard. They are also willing to give you limited advice on how to comply with it. The registrar is not allowed to do either during the final assessment. The outcome of a pre-assessment, however, is the same as for the final audit. The registrar will list all the observed deficiencies and appraise your readiness for the certification audit.

During this process at Mountaintop, the team for each section met with the registrar. They experienced the style of audit, the pressure of having to have all the answers, and the feedback of the interpretations. This prepared them for the next stage. To say that the pre-assessment was worth every penny is a significant understatement.

Final Preparation

The pre-assessment provided valuable experience and exposed some deficiencies in the quality system. The steering committee met with the teams to map out the strategy for addressing the deficiencies. Three major areas required system improvements and several needed minor corrections. The existing teams addressed the minor areas. New teams were formed with representatives from all areas of the system to correct the major system issues.

The ISO 9000 series standards usually require at least two to three months of evidence of compliance. The system corrections were the hardest to implement in time to meet this requirement. Meetings took place at least twice a week. The steering committee reviewed the progress on a biweekly basis. The pre-assessment experience also helped internal audit teams conduct several much-improved audits during this period.

Finally, the site updated many sections of the quality manual and second-tier documents to reflect the teams' recommendations and changes. (Figure 10.3 shows the document hierarchy.) Again, the site had to communicate these changes to all levels of the organization. The plant's staff managers and the management ISO representative did this in several communications meetings. Chapter 5, on teaming to win, discusses the site's monthly communication meeting.

The Certification Assessment

The day arrived for the five-day certification audit that would test the entire organization's efforts. The registrar's team consisted of two assessors, including one lead assessor. The lead assessor was the same individual who conducted the pre-assessment. Harris has found it very helpful to maintain continuity by using the same assessor on all periodic (follow-up) audits.

Individual assessors met with the teams and areas responsible for each section. It was vital, however, that at least one member of the steering committee always accompany each assessor. This avoided controversy about the findings (deficiencies) noted and ensured continuity for follow-up actions. At the end of each day, the assessors presented their findings to the employees. The steering committee organized whatever supporting information the assessors needed for the next day. The fifth day consisted of closed-door session with the assessment team, which presented its findings to management.

There were no class I (major) findings, and only a few class II deficiencies. Mountaintop addressed all of them within a 90-day period. The assessment team felt the corrective actions were sufficient without the need for a follow-up verification. DNV issued the certificate on August 19, 1994.

Just the Beginning

Many companies let up on themselves after the one- to two-year intensive preparation for certification. Even the best foundation needs periodic reinforcement. The program is always subject to breakdown because new employees did not live through the process.

The only way to reduce this effect is to make ISO 9000 and QS-9000 part of the organizational culture. Every employee must believe that ISO 9000 is not merely the "quality program of the year (or month)." The idea of management commitment to quality programs recurs throughout this book. Harris has taken several steps to reinforce this sense of commitment.

1. Employees designed and implemented most of the system.

2. The bimonthly engineering/management review meeting discusses and documents various elements of the program. It is part of the monthly all-employee communications meetings.

3. The Harris quality policy and the plant's goals appear on the back of employee identification badges.

4. The internal audit team is active all year. It includes members of all disciplines, including secretaries. The members rotate annually, so the knowledge core is constantly growing throughout the plant.

5. All new employees receive training in ISO 9000.

6. All levels of management constantly challenge their departments about ISO compliance when system changes are proposed.

The evidence is clear. Production operators frequently challenge new procedures with the question, "Does this comply with ISO standards?"

QS-9000: The Next Step

The Automotive Companies Develop a Common Quality System

In September 1994, a task force of the North American automobile companies released *Quality System Requirements: QS-9000*. It is a common, worldwide set of supplier quality requirements for Chrysler Corporation, Ford Motor Company, and General Motors Corporation. Its basis is the 1994 revision of ISO 9001, and it includes the text of that standard.

QS-9000 added 100 new requirements to cover additional areas that are particular to the automotive industry.

QS-9000 has three major sections.

1. Section 1, "ISO 9000-Based Requirements"

2. Section 2, "Sector-Specific Requirements"—this covers the production part approval process (PPAP), continuous improvement, and manufacturing capabilities

3. Section 3, "Customer-Specific Requirements"—this section lists requirements for which each of the Big Three maintains separate practices

Chrysler, Ford, and General Motors require their tier 1 suppliers of production and service parts and materials to receive second-party (AEC) or third-party certification. (AEC is the Automotive Electronics Council.) Many tier 1 suppliers require third party certification of the tier 2 supply base. (A first-party audit is a self-assessment, and a second-party audit is an audit by a customer. A third-party audit is an assessment by an independent party, often a registrar.)

QS-9000 Includes the Entire Flow

QS-9000 requires ISO 9001 compliance. A manufacturing plant can certify to ISO 9002 if no design activities are conducted on site. ISO 9001 adds requirements for product design. QS-9000 includes the entire flow through design, marketing, and customer support. It usually involves all sites involved in the process, even if they are located outside the United States. An organization that meets QS-9000 requirements also qualifies for ISO 9000, but ISO 9000 compliance does not ensure readiness for QS-9000.

Additional Emphasis

QS-9000 includes more than 100 additional requirements beyond ISO 9000. There also are minor clarifications and enhancements to each section. It is far more data driven and more demanding. However, QS-9000's major demands cover continuous improvement, business planning and practices, prevention, and statistical systems and tools.

Additional Tools and Systems

Some of these additional areas are logical extensions of ISO 9000. Others, however, are unique to the automotive segment, or involve structured implementation of certain tools.

Failure Modes Effects Analysis. Failure modes effects analysis (FMEA) is a procedure for risk analysis. From the Automotive Industry Action Group FMEA manual (AIAG 1995),

> *An FMEA can be described as a systematized group of activities intended to:*
>
> *1. recognize and evaluate the potential failure of a product/process and its effects.*
>
> *2. identify actions which could eliminate or reduce the chance of the potential failure occurring, and*
>
> *3. document the process.*

Brainstorming, with participation by the affected engineering and manufacturing parties, is the best procedure. FMEA should happen as early as possible in the design and manufacturing cycle. It may, however, be necessary to recapture past activities to meet the standard. QS-9000 requires formal documentation of this analysis for all designs, and all processes, including purchased materials.

Production Part Approval Process. According to the AIAG PPAP manual (AIAG 1995), "The purpose of production part approval is to determine if all customer engineering design record and specification requirements are properly understood, and that the process has the potential to produce product meeting these requirements during an actual production run." Suppose a supplier wants to make a design or process change on a previously approved part, or any new product. The supplier must submit specific forms and information to its QS-9000 customers before doing this.

Control Plan. The AIAG Advanced Product Quality Planning and Control Plan manual (AIAG 1995) defines a control plan as "A written summary description of the systems used in minimizing process and product variation." It is a "living document, reflecting the current methods of control, and measurement systems used." This is a logical subset of the quality planning section of ISO 9000, but the format and content are very specific.

Statistical Process Control. ISO 9000 does not require the use of statistical techniques to control the manufacturing processes. (Section 4.20 does apply to these techniques, however, if they are used.) QS-9000 not only requires statistical techniques, but sets standards for process capability (C_{pk} and P_{pk}). It also sets standards for measurement system capability (gage reproducibility and repeatability). Chapter 12, on statistical methods, discusses process capability and gage capability. These elements have direct ties to the control plan.

Continuous Improvement. This is the most data-driven cornerstone of the automotive standard. Every facet of the business must have goals for improvement, from purchased parts price reductions to 100 percent on-time delivery. The QS-9000 requirements assume there are goals and measurements for everything important to one's business. The auditors will, however, require simple trend charts for at least a year's performance for each measurement.

Here are some continuous improvement activities that Mountaintop looks for during internal audits.

1. Build or control quality into the product, instead of trying to inspect or test it in.

2. The minimum standard for the C_{pk} process capability index is 1.33. Processes with lesser capabilities require action plans for improvement. We want 1.67 or better, since this ensures less than 0.6 ppm nonconformances. Chapter 12, on statistical methods, shows that no realistic attribute (good/bad) sampling plan can ensure 0.6 ppm quality.

3. Improve productivity and quality. Chapter 8, on total productive maintenance, discusses ways to improve equipment efficiency. Chapter 7, on zero scrap actions, discusses how teams reduce scrap and rework. Synchronous flow manufacturing improves productivity, while statistical methods help improve the process.

Again, Mountaintop's quality programs reinforce and support each other. Here, they support the continuous improvement goals of QS-9000.

Certification Process

For Chrysler and General Motors suppliers, certification must be through third-party audits by qualified ISO 9001 registrars. Ford accepts second-party audits (any or all of the AEC member's supplier quality organizations) as evidence of compliance. Ford also accepts third-party audits by qualified ISO 9001 registrars. However, not all registrars for the ISO 9000 series standards are qualified to conduct third-party audits to QS-9000. A list of recognized accreditation bodies is available at ASQ's QS-9000 web site (http://www.asq.org/9000/).

Mountaintop Prepares for Its Compliance Audit

The automotive market accounts for more than 30 percent of Harris' power semiconductor sales. With ISO 9002 certification complete, the plant started to prepare for a second-party QS-9000 audit. The audit team would consist of representatives from each of the AEC members. It was this heavy involvement that dictated the second-party registration. No third-party registrars were then certified. We do not recommend this approach today, since third-party registrars are available. The major areas of concern were design and marketing, since these areas were not part of the ISO 9002 requirement.

Design. Unlike the manufacturing organization, neither design nor marketing had ever undergone auditing to any standard. The plant assigned a full-time person to coordinate the design activity. This individual put into action the team concept that had worked so well for the ISO 9002 preparation. The team concept worked so well that it replaced the traditional top-down organization permanently.

Manufacturing had already been applying most of the elements because of its large automotive market. Therefore, the quality manager and the ISO 9002 management representative provided insight and expertise about both standards to the design organization. We added internal auditors from the design group to the existing ISO 9000 team, which immediately conducted a pre-assessment audit. We also decided to integrate design into the ISO 9002 quality manual, which avoided duplication.

The existing documentation system for design required each engineer to maintain all the records for his or her designs. The standard specifies records for the design innovation process, design reviews, verification, and transfer to manufacturing. It became obvious that a computerized system on a local area network was the best way to give each team member access in each phase. It would also provide organized record retention. The entire system would use standard spreadsheet and word processing software.

The entry of the engineering laboratories into the calibration system was a parallel activity. The laboratories also had to comply with the AIAG measurement system analysis reference manual (AIAG 1995). Although all design specifications were in the formal documentation system, there were no formal instructions for the design process. There were no detailed instructions for operations like the computer-aided design tools, either. Design teams developed instructions for both applications.

Marketing. The marketing organization is not in Mountaintop and is not part of any manufacturing plant. It was therefore necessary to create a separate quality manual and institute all the applicable quality systems. The process was similar, but not as extensive as for the manufacturing plants.

Manufacturing. The manufacturing operation had supplied the automotive industry for 25 years. Many of the systems were therefore already in place. The major effort for compliance was to document them to the letter of the standard. We also had to present the supporting data for continuous improvement, competitive benchmarking, and business planning practices in concise charts and tables. The same teams and steering committee that had successfully organized the ISO 9002 sections received this task. The internal audit team had to learn the QS-9000 requirements thoroughly and had to add many new sections and requirements to the quality system documentation.

The Benefits

In May 1996, the Mountaintop plant was issued a certificate of compliance to QS-9000 through a second-party assessment by all members of the AEC. Mountaintop will soon seek third-party certification through its ISO 9001 registrar. DNV has passed the qualification process administered by a recognized QS-9000 accreditation body.

The three automotive companies of North America developed the QS-9000 standard. This has made it much simpler for suppliers to meet their quality requirements. Suppliers formerly had to meet each customer's requirements, and each customer performed its own compliance audits. ISO 9000 has not yet achieved comparable standardization, since many other companies do not feel that it covers all their needs adequately. However, the biggest benefit to Mountaintop was the integration of design and marketing into the team and quality system culture.

Calibration

Calibration is among the most important elements of a quality system. It is tempting to underemphasize it, since it apparently adds no value to the process. Its role in quality assurance is, however, critical.

Robert M. Bakker of Entala, Inc. (Grand Rapids, Michigan) cited the following elements of QS-9000 as major problems for many companies (Bakker 1996).

1. Documentation control (element 4.5): "Little yellow sticky notes with work instructions stuck to documents or machinery won't cut it"

2. Inspection and testing (element 4.10)

3. Control of inspection, measuring, and test equipment (element 4.11): "There's often at least one gage with no record of calibration"

We can look at Bakker's observations from a Pareto Principle standpoint. The Pareto Principle says that most of the trouble (with anything) comes from a few sources. The general rule is that 80 percent of the trouble comes from 20 percent of the sources. We therefore expect most ISO 9000 nonconformances to involve documentation, inspection and testing, and calibration.

Chapter 12, on statistical methods, discusses statistical process control (SPC). SPC also must rely on accurate measurements. The calibration system, therefore, supports SPC.

Calibration and Quality

Calibration plays a key role in assuring product quality. The factory must rely on measurements and parametric evaluations at various stages of the process. Calibration ensures these measurements' accuracy, reliability, and validity.

Consider a simple consumer product like an incandescent light bulb. The manufacturer publishes the bulb's wattage and expected lifetime, and consumers rely on this information. Buyers use it to judge the bulb's utility for a given application. The manufacturer presumably samples bulbs and measures their wattage. If the power meter is not accurate, neither the manufacturer nor the consumer can rely on the specification.

Here is another example. An expensive fish or meat costs $8.00 a pound, or 50 cents an ounce. If you are buying this product, you want some assurance that the vendor's scale is accurate. Suppose you buy it every week. If the vendor's scale adds half an ounce to the weight of each purchase, it will cost you $26.00 a year. This is why there are special regulations for scales, and why spring scales are "not legal for trade." States and municipalities have departments of weights and measures that calibrate scales routinely. Similarly, gasoline pumps are subject to regular calibration.

Calibration Program: Goals

A calibration system's principal goal is to ensure the relationship between an instrument's measurements and a standard. If a standard is exactly 0.10 inch (2.54 mm) wide, measurements from a calibrated micrometer will average 0.10 inch.

The words "average 0.10 inch" introduce the difference between calibration and capability. Few gages or instruments will report exactly the same number every time the user measures the specimen. There is inherent variation in the measurements, which comes from (lack of) reproducibility and repeatability. If the measurement depends on the operator, this is a lack of reproducibility. If the same operator gets different measurements from a single specimen, this is lack of repeatability. A gage capability study, or reproducibility and repeatability (R&R) study, quantifies this variation.

A gage study reports the percent tolerance consumed by (lack of) capability, or PTCC. A PTCC of less than 10 percent is good, and QS-9000 calls for this. A gage whose PTCC is more than 30 percent is not capable. Even if the gage is accurate, there is so much variation in its measurements that we cannot rely on them. A noncapable gage raises the chances of shipping nonconforming pieces and rejecting good ones. Figure 10.4 is a contour plot of gage response versus actual dimension (Levinson and

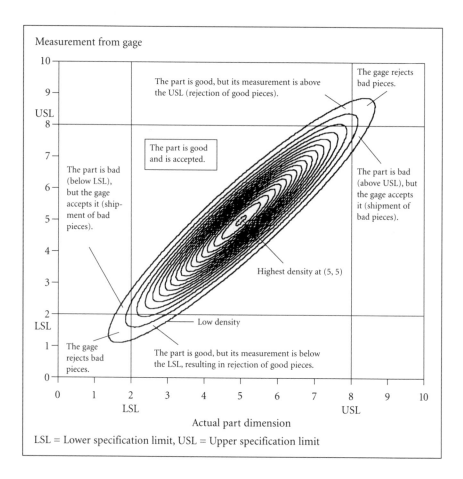

Measurement from gage

The part is good, but its measurement is above the USL (rejection of good pieces).

The gage rejects bad pieces.

The part is good and is accepted.

The part is bad (below LSL), but the gage accepts it (shipment of bad pieces).

The part is bad (above USL), but the gage accepts it (shipment of bad pieces).

Highest density at (5, 5)

Low density

The gage rejects bad pieces.

The part is good, but its measurement is below the LSL, resulting in rejection of good pieces.

Actual part dimension

LSL = Lower specification limit, USL = Upper specification limit

Figure 10.4. Gage capability and outgoing quality ($C_p = 2/3$, PTCC = 50%).

Tumbelty, 1997, 175). When the product's dimension is borderline, the gage can accept bad pieces or reject good ones.

The gage's accuracy, however, does not depend on its capability (see Figure 10.5; see also Levinson and Tumbelty 1997, 151). An accurate gage will, on average, report the standard's correct measurement. An inaccurate or out-of-calibration gage will not. A gage may be noncapable but accurate, or capable but inaccurate. We obviously want our gages and instruments to be both capable and accurate.

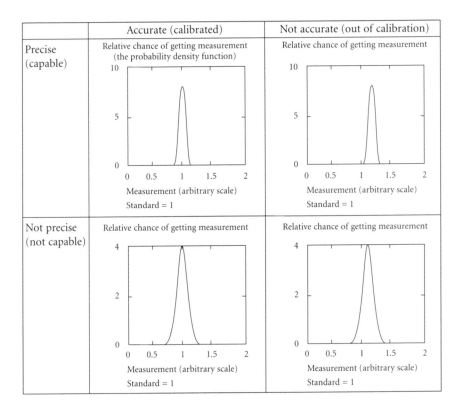

Figure 10.5. Gage accuracy and precision.

The Calibration Laboratory and Environmental Controls

There are often environmental specifications for calibration, because temperature and humidity may affect the standard or gage. For example, you are better off buying cold gasoline than warm gasoline. You pay by the gallon or liter, but cold gasoline is denser than warm fuel. Gases are often metered (or sold) by standard cubic feet. (One cubic foot equals 0.02832 cubic meter.) The "standard" refers to standard conditions of one atmosphere (1.01325 bar, or 14.696 psia) and 32°F (0°C). The environmental specification is important because a gas' density depends on the temperature and pressure. Temperature and humidity may affect other measurements, too. A calibration lab often requires strict environmental controls.

Calibration and Standards

Calibration compares the instrument's or gage's measurement against a standard, often under specific environmental conditions. Consider the incandescent light bulb again. Power, or wattage, is potential (volts) times current (amperage). The manufacturer must measure the current that the lamp draws and the voltage.

The standard could be a voltmeter with an accuracy of 0.01 percent, while the factory gage's accuracy is 1 percent. To calibrate the factory gage, one would connect both instruments to a single voltage source. If the factory gage is accurate, it will report the same voltage as the standard voltmeter, ±1 percent.

The accuracy ratio between the standard and the instrument under calibration must be at least four to one. The standard must, in other words, be at least four times as accurate as the factory gage. This is the minimum nominal relationship for an acceptable calibration. In the example, the ratio is 100:1.

Calibration Schedules

Accuracy over time is another vital part of the calibration program. How long does a calibrated instrument stay within limits? Gages and instruments drift out of calibration primarily through usage, but also with time (Juran and Gryna 1988, 18.78). The calibration program must define a calibration schedule for each instrument and gage. The schedule must be rigorous enough to detect and correct deterioration in the gages' accuracy before the deterioration exceeds acceptable limits.

There are several ways to quantify a gage's or instrument's usage (Juran and Gryna 1988, 18.78).

1. Elapsed time—this is the most common method. For example, a schedule may call for recalibration every six months.

2. Actual usage, in units of product.

3. Actual operating hours, primarily for electrical equipment. The factory meters the time during which the instrument is drawing power.

4. Test accuracy ratio control.

Harris' Mountaintop plant relies primarily on the elapsed time method.

The gage's manufacturer may, of course, specify or recommend a schedule. If not, there are several algorithms and statistical methods for finding an appropriate interval. If the gage meets a series of in-tolerance conditions at each successive check, we can increase the interval between checks. If there is a series of out-of-tolerance conditions at successive checks, the calibration interval decreases.

Traceability of the Instrument Chain

All calibrations rely on traceability to a chain of instruments. Every instrument that calibrates another must have an unbroken link to a common, accurate reference. The end (or top) of this traceability chain is usually a nationally or internationally recognized standard.

In the United States, the traceability chain begins at the National Institute of Standards and Technology (NIST) in Fort Collins, Colorado.* The NIST umbrella includes standard references for everything from linear dimensions to time and energy. NIST has accepted or assigned values for fundamental or natural physical constants. These include the speed of light in a vacuum (c) and the quantized Hall resistance (R_H).

Correlation

There are a few very limited situations in which there are no true standards. In these cases, suppliers and customers have set up de facto standards by mutual consent. When suppliers and customers agree on a reference, the situation involves correlation. Correlation systems allow determination of an instrument's or system's repeatability. Correlation is similar to calibration, but there is no traceability to an NIST standard.

A correlation chain usually has only one link to the mutually agreed reference. The system's main validity comes from a long history of repeatability. Correlation's main disadvantage is its lack of an accurate way to trace measurements to true standards. It is not an acceptable substitute for calibration if traceable references are readily available.

---

*http://www.nist.gov/; 325 Broadway, Boulder, CO 80303; Public Inquiries Unit, 301-975-3058.

Calibration: The Rules

A calibration system includes schedules, traceable instruments and standards, and documentation. There is a set of rules for maintaining the system's integrity. Most calibration departments follow ANSI/NCSL Z540-1-1994. ANSI is the American National Standards Institute,* and NCSL is the National Conference of Standards Laboratories. This standard is a compilation of many documents, procedures, and publications. The oldest include basic laboratory procedures from the nineteenth century. Others are more formal, like basic calibration methods from Volume 5 of the *Electrical Engineers' Handbook* (Pender and McIlwain 1936). The most recent of these historic documents was MIL-STD 45662 and MIL-STD 45662A. It covers setup and maintenance of calibration systems and the control of measurement equipment and standards. Its purpose was to ensure that services and supplies for the government meet contractual technical requirements.

Evolution of Standards

The U.S. government canceled the military standards in January 1995. Although the government will no longer update the document, it is a useful guide for basic calibration systems. The newest procedural documents use the same methods, but they encompass the trend toward a global environment.

The principal goal of the MIL-STD was to assure that goods and services for the U.S. government met technical requirements. The standard also created a framework for systems to control calibration of any group of measurement and test equipment. By doing this, it helps assure the quality of goods from equipment that falls under the system's control. The control methods go beyond ensuring the accuracy of the products' parametric values by comparing them with a standard. They also promote repeatability between product runs. The overall goal is to ensure that products from a manufacturing system meet the customer's requirements.

---

*http://www.ansi.org; American National Standards Institute, 11 West 42nd Street, New York, NY 10036; Telephone 212-642-4900; Fax 212-398-0023.

Manufacturers adopted the original military standard as a normal method of doing business. When this happened, the government no longer needed to maintain the original standard. This is why the government canceled the standard in 1995 and suggested that vendors use the new ANSI/NCSL Z540-1-1994 standard. This new standard included all the features of the MIL-STD and added detailed requirements for laboratories that provide calibration services. The new standard also incorporated the ISO 9000 guidelines for calibration programs. These additions aligned calibration systems with the international marketplace and made it easier to trade there.

The most important additions deal with quality issues. The implications of a quality-relevant service placed a new emphasis on the services that the calibration service's customer requires. Calibration customer satisfaction is the final goal of an entire calibration operation. This, in turn, helps ensure that products from the calibrated equipment will meet the end user's needs.

The Practice

Every manufacturing operation must control whatever process it uses to produce its goods or services. It must therefore be able to measure all the parameters it uses to control the operation. Measurements for even simple products can be very complex.

Consider again the incandescent light bulb. Its glass envelope requires several parametric evaluations. The proportions of sand and other materials for the glass must meet tight specifications. The temperature for the glass blowing operation is another critical parameter. The weighing and temperature measurement equipment therefore need calibration control. After glass blowing, the envelope must cool at a specific rate to avoid stresses that could lead to breakage. After the envelope has cooled, the thickness of the glass must be measured to ensure structural integrity and repeatability in the process. The list can be very extensive for what seem like simple products.

Measurements are key to all manufacturing operations, and they must be accurate to ensure process control. When the process dictates the measurement's accuracy, the instrument's calibration is mandatory. A very accurate gage makes it easier to control the process. Calibration therefore supports SPC, design of experiments (DOE), and other quality control tools that rely on accurate measurements.

Also remember the importance of gage capability: The gage must be precise as well as accurate. SPC and DOE will work if the gage is inaccurate, since the gage introduces the same error into every measurement.* If the gage drifts out of calibration, however, it will undermine these activities. Gage variation, or percent tolerance consumed by (lack of) capability, reduces the power of SPC and DOE. A test's power is its ability to detect a process shift or a difference between treatments.

Attitudes Toward Calibration. Manufacturing organizations have viewed calibration as a necessary evil (see Figure 10.6). It costs money, and it adds no value to the product. Chapter 9, on synchronous flow manufacturing, discusses the dysfunctional effects of dogmatic cost accounting systems. These systems look solely at the bottom line, and they guide business decisions accordingly. They don't even look at intangibles such as customer satisfaction!

This attitude was even stronger in the early part of the twentieth century, when consumers looked at an item's price and little else. It began to change as people started counting the cost of getting rid of and replacing poor-quality goods. The ever-expanding scrap automobile yards were a prime example. Customers began looking closely at products' durability and quality and began buying accordingly. This is why many people started buying Japanese automobiles during the 1980s.

American manufacturers realized that they had to improve their quality to remain competitive. They soon discovered that quality assurance relies on measurement accuracy, and calibration gained a new respectability. Manufacturers realized that calibration helped them reduce waste and improve process yield. Less waste and higher yields, in turn, meant higher

---

*In SPC, suppose that all the measurements are 0.1 micron too low. Each sample average will be 0.1 micron too low, but so will the control chart's center line and control limits. (This assumes that we have based them on data from the out-of-calibration gage.) The control chart will therefore perform its job properly. The systematic error will, however, compromise the outgoing quality. If the upper specification is 18.0 microns, everything we think is in the range [17.9, 18.0] microns is actually nonconforming.

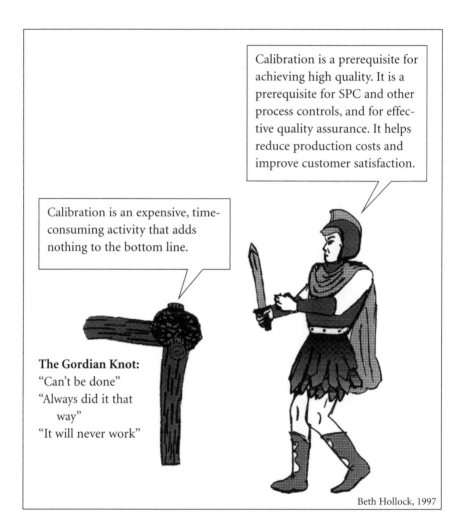

Figure 10.6. Attitudes toward calibration.

profits. The products performed better, lasted longer, cost less to operate, and satisfied the customers.

Calibration and Quality. Manufacturers started to look even more closely at product quality. Statistical tools showed that calibration control improves product quality. This created an expanding demand for calibration services. The same philosophies that helped with the manufacturing operation integrated themselves into calibration. The results of a well-tuned calibration system carried through to the end user. When process control ensures quality, customers don't have to worry about the product. Customer satisfaction surveys quantify how well a manufacturer's quality performs against the competition. Much of the quality improvement was traceable to the calibration activity. These improvements affected both domestic and international sales.

The ISO 9000 series standards require a calibration program, as do most quality management systems. Calibration activity is vital for achieving world-class manufacturing.

Calibration and Customer Satisfaction. We cannot underemphasize the importance of customer satisfaction. Struebing (1996) says that it costs five to seven times as much to get a new customer as it does to satisfy and keep one. We are better off spending $1000 on quality assurance (including calibration) than $5000 on advertising and promotion. Suppose $200,000 will operate a calibration lab for a year. This is like spending a million dollars for one TV ad during the Super Bowl.

This assessment does not even consider customer advocacy. Harris Semiconductor looks at both repeat business and customer advocacy. Mountaintop asks its customers, "Would you buy from us again?" and "Would you recommend our products?" Happy customers recommend our products or services to others, which is free advertising.

Unhappy customers, especially consumers, badmouth the offending products and their manufacturers. There is even a Web site for "hate mail" about businesses. The page's owner is indexing writers' negative experiences by company. This is extremely dangerous, because search engines that catalog this site use the company names as key words. An Internet user who searches for "Brand X" may find the hate mail Web site before finding Brand X's! Chapter 13, on the Internet, projects the Internet's influence on business in the twenty-first century. Calibration, and other quality assurance techniques, are better investments now than ever.

The Real-Life Story: Calibration at Harris

The Mountaintop plant experienced changes in the competitive environment and attitudes toward calibration. When the RCA Solid State Division owned the plant, management viewed calibration as a costly necessity. The plant provided as little staffing and equipment as possible. This practice was not deficient under contemporary standards; the customers were happy. Customers audited the plant, but did not consider this a problem. General Electric's takeover of the plant did not lead to any major changes, either. Customers still found the situation acceptable.

Harris' acquisition of the plant led to some minor changes. During the late 1980s and early 1990s, however, many companies were looking at trends in the European marketplace. Some of Harris' customers were among these companies. Conformance with ISO 9000 was becoming a condition of doing business in Europe. Companies were adopting the standards and placing greater demands on their suppliers' quality. Harris realized that it had to assure quality through process control, and this placed greater demands on the calibration laboratory. The Mountaintop plant was also looking at the requirements for ISO 9000, but was not yet seeking certification.

The audit requirements of customers who were active members of the ISO program forced a careful reexamination of the calibration activity. The plant selected a member of the engineering community to drive the activities of the laboratory. To revitalize the program, the plant shortened the chain between the instruments under calibration control and NIST. Newer and more accurate standards replaced some very old ones. This program improved the accuracy ratio for many links in the calibration chain.

When Harris became an active participant in the ISO certification program, calibration assumed a new importance. The plant added staff to the laboratory to help meet these needs. Meanwhile, new markets were prompting the development of new products. These new products, in turn, demanded new metrology methods and equipment. The plant had to invest capital to improve the laboratory's ability to handle these new requirements. Mountaintop also launched a program to make manufacturing equipment end users more active contributors to the calibration activity. In the new culture, equipment operators frequently ask about specific tools on the calibration schedule. The operators appreciate the program's importance and want to make sure that their instruments are properly calibrated.

The recent improvements in staffing and equipment have given Mountaintop's calibration program world-class capability. Other manufacturers, including Harris' customers, are now benchmarking themselves against it. They are using it to map their own journey toward ISO 9000 certification.

CHAPTER 11

Integrated Yield Management

Robert Fitch and Steve Tetlak

Mountaintop's integrated yield management (IYM) program is the catalyst for a paradigm shift in yield thinking. The entire design and manufacturing pipeline's yield is the critical metric of the plant's performance.

Each manufacturing entity in the pipeline was once responsible for its local yield (see Figure 11.1). The product's final yield was not a consideration for the pipeline's beginning stages. New attitudes encourage each manufacturing area to accept ownership of final yield. The new methods require extensive communication between areas and integrated data management.

The design activity's impact on yield is recognizable as design for manufacture (DFM). Meanwhile, how can an operation affect the yields of those downstream? Readers in non-semiconductor industries are probably familiar with additive tolerances. Suppose that three parts must nest or stack in the final assembly. The final result can be out of specification, even if the three parts are within theirs. The problem in the semiconductor industry is similar.

Yield, Throughput Yield, and Pipeline Yield

Yield is easy to understand. It's the ratio of good product to starting material. Semiconductor manufacturers process electrical devices in wafer form until final testing and dicing, and there are two yields. The throughput yield (TPY) is the fraction of wafers that survive to final test. The final

Every operation's performance can affect the overall yield. Furthermore, product design affects yield. Departments must communicate and work together for high overall yields.

High yields at each operation translate into high overall yields. We can measure each operation's performance in isolation.

The Gordian Knot:
"Can't be done"
"Always did it that
    way"
"It will never work"

Beth Hollock, 1997

Figure 11.1. Yield paradigm.

test (FT) yield is the fraction of the electrical devices (chips, die, or pellets) that are good. The overall yield is the product of the throughput yield and final test yield.

Chapter 9, on synchronous flow manufacturing, shows that yield becomes critical in a constrained process. Process yield losses after the constraint are irrecoverable and irreplaceable; the factory does not have the capacity to replace the losses. Imperfect test yields also are irrecoverable, because the factory cannot replace them. In the engineering/managerial economics world, these translate into opportunity costs. The forgone opportunities to earn revenue can be far more expensive than what the cost accounting system shows.

Pipeline Yields in Multistep Processes

Yield is particularly challenging in long multistep processes, which are typical of the semiconductor industry. Consider the following scenario. A small wafer fabrication (fab), test, and assembly plant has five divisions (see Table 11.1 and Figure 11.2).

Table 11.1. The wafer manufacturing multistep process.

Division	Description (may be skipped without loss of continuity)	Abbreviation
Epitaxial silicon growth	Deposition of crystalline silicon on a wafer. The epitaxial silicon contains a dopant that gives it electrical properties.	EPI
Wafer fabrication facility	Implantation of dopants into the wafer to create microscopic transistors and connection of these electrical devices with metal wiring.	FAB
Wafer testing facility	Testing of the electrical devices (chips, die, or pellets) and rejection of those that do not meet specifications.	PROBE
Silicon die packaging assembly facility	The electrical devices are mounted on stems and equipped with protective metal caps or plastic encapsulations. The product now looks like a familiar transistor.	ASSY
Final test facility	The packaged devices are tested to make sure they still meet specifications. Also, this operation can do high amperage tests that the wafer tester cannot perform. The wafer tester's wires must be thin to probe the tiny chips, and they cannot carry a lot of power. Packaged devices can accept full operating loads.	FT

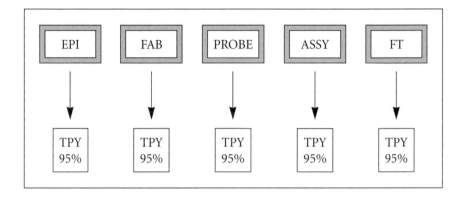

Figure 11.2. Throughput yield of each division within the plant.

There are six basic steps in wafer fabrication.

1. Photolithography, or definition of microscopic features on the wafer, is similar to photography. The first step is to coat the wafer with a photosensitive polymer film, or photoresist. An exposure tool, which is like a giant camera, exposes the film to a mask with the desired pattern. The exposed photoresist dissolves in the developer.*

2. Ion implantation drives dopant atoms into the silicon, which creates the electrical devices.

3. Diffusion spreads the dopant atoms within the silicon. At high temperatures, atoms will diffuse even in solids.

4. Metal deposition puts a layer of metal over the wafer. After photolithography and etching, this metal will become the wiring for the electrical devices.

5. Another photolithography step defines the metal wiring pattern.

---

*A positive photoresist becomes soluble after exposure. A negative photoresist is soluble and becomes insoluble after exposure. Most semiconductor fabs use positive resists.

6. An etchant removes exposed metal and leaves a wiring pattern behind.

After fabrication, the complete wafer goes to an electrical probe that tests each device for proper performance. The tester probes every unit and marks the bad units with ink dots. The ink dots are recognizable by picking equipment, which rejects the bad units. A diamond-edged saw cuts the wafer into individual die. An automatic picker then selects the good ones for packaging and shipment.

The electrical devices are small, and their electrical connection points are tiny. The probe tips have to be equally small, which limits the power they can deliver to the devices. The products are designed to carry 50 amps, or even more, so the wafer probe cannot test the devices completely. After packaging, the devices receive a full power test before shipment (Figure 11.3).

Yields in Multistep Processes. Figure 11.2 shows the basic divisions of this manufacturing process. Each block (or section of pipe) is a part of the entire manufacturing pipeline. Each block has its own throughput yield

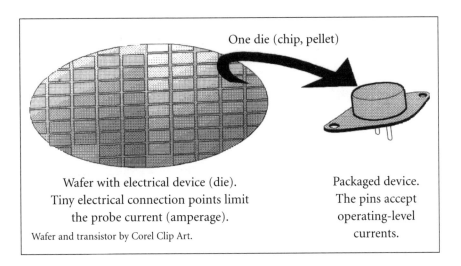

One die (chip, pellet)

Wafer with electrical device (die). Tiny electrical connection points limit the probe current (amperage).

Wafer and transistor by Corel Clip Art.

Packaged device. The pins accept operating-level currents.

Figure 11.3. Wafer and packaged device.

(TPY), as noted by the TPY box in the figure. Throughput yield (in percent) is

$$TPY = \frac{\text{Wafers out}}{\text{Wafers in}} \times 100\%$$

If operation i's yield is y_i (expressed as a fraction),

(Eq. set 11.1)

$$TPY = \prod_{i=1}^{n} y_i \text{ for } n \text{ operations}$$

A 95 percent yield for each block or division sounds great. A close look at Equation Set 11.1, however, should give us an uneasy feeling. If $y = 0.95$ for each block and there are five blocks, $TPY = 0.95^5 = 0.774$, or only 77.4 percent.

Wafers at Mountaintop actually pass through many operations. Also, processes for power transistors are simpler than those for computer chips. A wafer with computer chips may pass through hundreds of operations. A 99 percent yield sounds very good. It's like playing Russian roulette with one bullet in 100 chambers, and the chance of surviving one round is 99 percent. The chance of surviving 100 rounds, however, is 36.6 percent. This is the TPY of a 100-operation process with a 99 percent yield at each step. Imperfect yields nickel-and-dime us to death in complex, multistep processes.

Parallel assembly is a way to avoid this. Suppose that there are 100 steps, and we can make four subassemblies with 25 operations each. If each operation's yield is 99 percent, the TPY for each subassembly is 77.8 percent. If the final assembly operation also has a 99 percent yield, the overall TPY is 77.8 percent times 0.99, or 77 percent. Chemists try to use this approach in multistep organic synthesis, and many manufacturing processes are amenable to it. There is, however, no way to fabricate silicon wafer "subassemblies" in parallel and put them together to make a complete wafer. The microelectronics industry has to live with long, multistep processes.

Example: TPY in a Multistep Process. The manufacturing pipeline in Figure 11.2 becomes a single entity, and the TPY of the pipeline depends on each segment's performance. Suppose we start with 100 wafers. These wafers enter the manufacturing pipe at EPI, which deposits epitaxial silicon on the wafers. One of the epi reactors malfunctions, and five wafers receive epi with the wrong resistivity. Epi resistivity, thickness, and number of light point defects are the three critical parameters for this process. EPI's TPY is therefore 95 percent, and 95 wafers leave EPI and enter the FAB.

How do the 95 wafers fare in the FAB division? Wafer transport occurs in a cassette. Each wafer goes in a grooved slot in the cassette, and there is about a quarter inch between wafers. Each cassette holds one lot of 10 wafers. (Lot sizes and wafer spacing will depend on the wafer diameter. At Mountaintop, these range from four to eight inches, or 100 to 200 mm.) The cassette goes inside a black plastic carrying box.

The box has a lot traveler (routing) with key information about the lot's identity: the part type, the lot number, the number of wafers in the box, the EPI parameters, and operators' initials showing when each operation was completed. Since EPI shipped 95 wafers, the FAB receives 10 lots, of which one has only five wafers. This FAB has six simplified silicon processing steps, which were outlined earlier. Figure 11.4 shows the first three steps.

Step 1 coats the wafers with photoresist. The wafers then go under a projection aligner that exposes the photoresist to a pattern on a photo mask. The deep ultraviolet light makes the exposed (negative) photoresist insoluble in a developer. The mask prevents light from reaching the areas we want to remove, and the developer dissolves the unexposed photoresist.

The photoresist pattern controls the next step: ion implantation. The implanter drives high-energy ions at the wafer, but the photoresist stops them. The dopant ions can enter the wafer only where the developer has removed the photoresist.

Next, a high-temperature furnace diffuses the implanted ions through the silicon to a precise depth and lateral width. This diffusion furnace reaches temperatures above 1100° C, so the wafers must go in a quartz boat. (The Teflon boat would not survive these conditions.)

A long time ago, operators had to use tweezers or vacuum pencils to transfer wafers between boats. Automatic wafer transfer tooling does away with human handling, but this equipment can sometimes malfunction. The transfer tool has not received proper maintenance, so it jams and breaks five wafers. Chapter 8, on total productive maintenance (TPM), describes preventive maintenance, which might have prevented

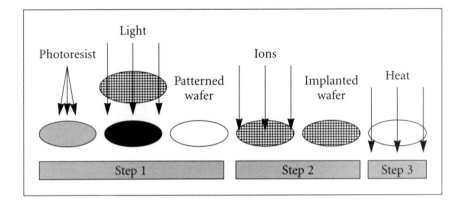

Figure 11.4. Fabrication processing: Photolithography, ion implantation, diffusion.

the breakage. TPM therefore goes beyond promoting tool availability and efficiency. It supports IYM by suppressing problems that can reduce throughput yield.

If there are no further losses in the remaining three steps, the FAB division's TPY is 94.7 percent. Round this off to 95 percent. Since EPI also lost five wafers, the pipeline is now down to 90 wafers.

The next division to receive the wafers is PROBE. This operation tests the quality of work produced from the previous divisions. Electrical parameter measurements and visual inspections screen out any failing die on the wafers. The electrical probe marks nonconforming die with ink, which prevents their use by the assembly process (ASSY). Also, wafers whose yields are too low are discarded. (We assume there is something wrong with them, and the problem may affect the apparently good die. For example, there could be a reliability problem that could lead to a field failure.) In PROBE, each lot yields 95 percent. Of every 100 die on each wafer, 95 are good. Since we have 90 wafers, we have the equivalent of 85.5 perfect wafers.

PROBE is, of course, not responsible for the five percent yield loss. PROBE is reporting a result that depends on the preceding operations. The throughput yields from EPI and FAB reflect wafer losses, but not die losses. The nonconforming die are bad, however, because of problems with the preceding operations. By the time the wafers reach PROBE, they have passed through the constraining operation. Chapter 9, on synchronous flow manufacturing, shows that bad die are an irrecoverable opportunity cost. It is therefore not enough to avoid losing wafers. The plant must also seek high probe yields on each wafer.

The assembly process cuts the wafers into individual die, then bonds these into plastic packages with external leads. These leads provide electrical connections to the die and the user's electrical circuit. If the assembly and final test yields also are 95 percent, the pipeline yield is only 77.4 percent. Assembly yields are high and FT yields are high too, but TPY methodology shows that our final pipeline yield is not what we expected. Figure 11.5 shows that only 77 wafers leave the pipeline. How could this have happened? While each division's yield looks good, the overall result is mediocre. The successive yield losses have nickeled and dimed us to death.

Management Information Systems and IYM

The factory tracks the wafers as they move through the pipeline. Harris uses Consilium's Workstream system to track the wafers. The system assigns a loss code to each wafer that becomes scrap. Loss codes include breakage by equipment, misprocessing, photoresist scumming, overetching, and other problems. Losses result from two general problem classes: breakage and misprocessing. Overetching, for example, might result from the wrong process conditions or bad materials. The loss codes help the engineers generate Pareto charts of losses from each area.

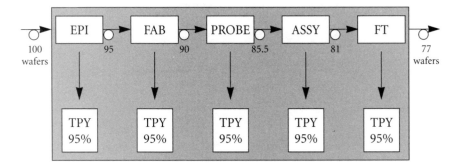

Figure 11.5. Individual yields and pipeline yield.

PROBE returns electrical test parameters, and these go into the Manufacturing Engineering Data Base (MEDB). The probe operation assigns a loss code to each defective die. The losses fall into bins. Typical bins for Harris' power transistors include the following:

- IDSS (drain-to-source shorts). *I* stands for current, as in $E = IR$. Read the acronym as "current, drain-to-source short." Current enters a field effect transistor's (FET's) source, passes through the gate, and exits through the drain (see Figure 11.6). The gate should control the current. (In a bipolar junction transistor, current enters the emitter, goes through the base, and leaves through the collector. The base controls the current.) A short circuit allows current to bypass the gate.

- IGSS (gate-to-source shorts).

- Vth (threshold voltage high or low).

- BVDSS (breakdown voltage drain-to-source high or low).

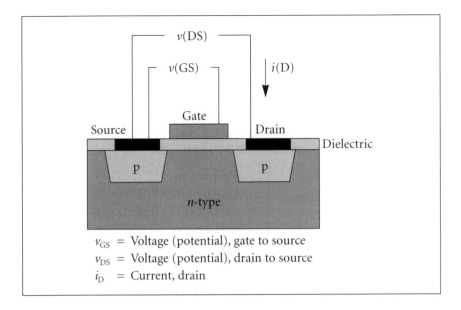

Figure 11.6. p-channel enhancement mode metal oxide semiconductor field effect transistor (MOSFET) (Smith 1976, 337).

The test equipment creates a list of test parameter measurements for each die and a list of bins for each wafer. Harris tracks these data with the MEDB. The losses at PROBE are also important to yield enhancement engineers. Each product typically runs at a certain yield level. When yields begin to vary, this shows that something may be awry in the FAB or EPI. A Pareto chart of the probe failure modes is vital feedback to the FAB. It allows identification and correction of manufacturing or process-related problems before larger losses can occur at PROBE.

Engineering data analysis of these losses is essential. Harris is developing two methods to track the probe yields: statistical bin analysis (SBA) and process average testing (PAT). The section later in this chapter on failure analysis covers these statistical tools and others. The latter include histograms, comparative box plots, and yield by lot or product.

Short- and Long-Loop Vehicles

The previous section used a mock facility as an example of a basic silicon manufacturing process flow. Within the flow, there were many process feedback loops. One was in EPI, and it monitors the epitaxial deposition temperature. This is an example of short-loop control. It is the best way to control an already established process. Statistical process control is also a short-loop vehicle that helps maintain a process' TPY.

The feedback from PROBE is a long-loop feedback vehicle. Probe data on product wafers ties the quality of the processed wafers back to the processing areas.

Long-loop feedback is rarely as effective as short-loop feedback (see Figure 11.7). It is a basic principle of process control that feedback loops should be as short as possible. As information becomes older, it loses its relevance to the process' current state. If we must wait until PROBE to detect a problem in EPI, the problem will have a chance to continue for a long time. Also, the EPI process may change well before its output reaches PROBE.

In chemical manufacturing processes, "dead time" and "transportation lag" refer to delays between processing and measurement. A stream from a heat exchanger may take some time to reach the thermometer or thermocouple. The controller for a plug flow reactor may act on the reactor's input, but it measures the output. The time delay equals the reactor's

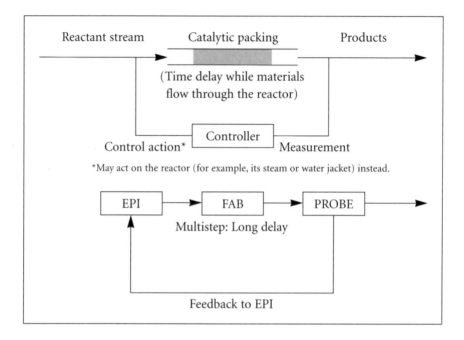

Figure 11.7. Long feedback loops.

residence time. Time delays between the process and the measurement destabilize control systems (Harriott 1964, 82).

The following section describes the basic IYM tools for both the short-loop and long-loop vehicles. They help maintain existing productivity and allow the plant to monitor productivity enhancements.

IYM Tools

Improvements as well as maintenance of existing yield levels require measurement. Within the IYM program, two important metrics are level limited yields (LLY) and defect density (DD). LLY relates to the FAB's TPY, and DD relates to PROBE yield.

Level Limited Yield

LLY provides ownership of yield losses to the responsible area in EPI or the FAB. For example, each operation in the FAB has a number in the processing sequence. The FAB includes the following areas: diffusion, etch, photo, implant, and metals. Each area is responsible for many operations, all of which periodically suffer wafer losses. Improper diffusion of the dopants can change the product's electrical properties, and make it fail at PROBE. LLY combines the TPY losses and the probe losses that are due to diffusion problems. Therefore, there must be feedback from probe to diffusion to identify the losses. Diffusion's LLY is defined as

$$\text{LLY}_{\text{DIFF}} = \frac{\text{Wafers out} - \text{Equivalent wafer probe loss by diffusion}}{\text{Wafers in}} \times 100\%$$

This definition assigns diffusion-related probe losses to the diffusion operation. Although the diffusion operation causes the scrap, it is not detectable before probe.

Each area in the fab as well as epi is responsible for increasing its level limited yield. Figure 11.8 shows loss tracking at each level. A Pareto chart helps each area identify key loss reasons and focus on improving its TPY.

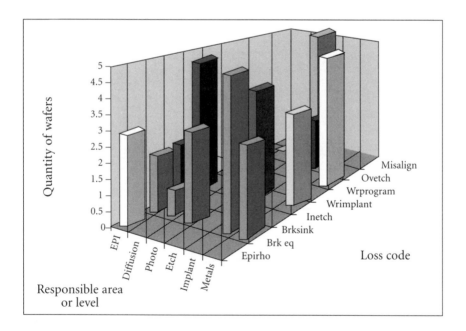

Figure 11.8. LLY loss tracking: Pareto chart.

Defect Density Metric

Yield is a good metric for probe results. Defect density is, however, a more meaningful metric for the quality of wafer output. DD measures the defect count per unit area. Harris Mountaintop uses the Berkeley Model to grade probe yields according to the following equation.

$$\%\text{YIELD} = [(1 - e^{-DD \times A})/DD \times A]^2 \times 100\%$$

where DD is defect density in defects per unit area and A is the area of the unit (die, chip, or pellet). Figure 11.9 shows yield versus die area for various die sizes. As the die size increases, the factory must find a way to reduce the defect density to maintain yields. The fabrication line's cleanliness has a strong influence on defect frequency. Tool cleanliness and cleanroom protocol (correct behavior and proper use of cleanroom garments) are extremely important here.

Actions to reduce defects on one product type can affect all products. The idea is to remove common cause (random) defect sources. Improvement involves suppression of common cause or random defects, while correction fixes special or assignable causes.

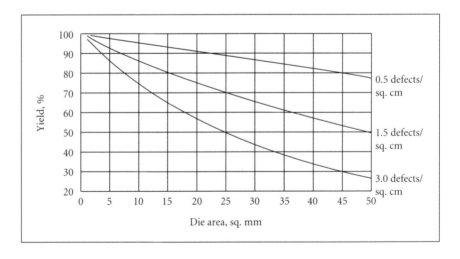

Figure 11.9. Yield versus die area and defect density.

Failure Analysis

The backbone of the Mountaintop IYM program is the systematic use of failure analysis. Priorities for failure analysis include the following:

1. Bad die on high yielding wafers (~90%). If most of the wafer is good, why are a few die bad? This analysis looks for systematic failure causes, which can affect several product types. The overall yield for high-yielding wafers is $Y = Y_S \times Y_R$, where S refers to systematic causes and R to random variation. Systematic losses result from causes that are inherent in the process. For example, the die at the edge of the wafer are rarely usable.

2. Zero-yielding wafers. These always result from special or assignable causes.

3. Whole wafer lots that escape from the fab and fail at probe. These are like zero-yielding wafers, but worse. They always result from special or assignable causes. Failure analysis can guide preventive actions to avoid future incidents.

This failure analysis is a long-loop feedback tool. Short-loop failure analysis is performed on wafers in the fab. These two represent most of the probe yield losses. The goal of failure analysis is to define 100% of the fab's losses to pinpoint key yield improvement projects. Accurate analysis of failing product provides essential feedback to the manufacturing areas.

To increase the plant's yield, we must understand why losses occur. In semiconductor manufacturing, there are two major forms of loss: scrap wafers and bad die. Wafer losses in the fab depress the TPY, while wafer losses at probe reduce the PT yield. The probe operation rejects individual die, and the final test rejects packaged die. (Remember that the probe cannot test the die at full operating power.) A Pareto chart of the rejection causes helps the wafer fab identify key yield issues. This Pareto chart comes from long-loop control (data from probe and final test). The information does not help control the process, but it helps the plant focus improvement activities.

Failure analysis begins with a thorough understanding of the electrical test data from probe. Three important tools help the yield enhancement engineer better understand these losses: wafer mapping, SBA, and

PAT. The next step is to locate the actual failure on the die, by liquid crystal and optical microscopy. (The liquid crystal makes it easy to find electrically induced hot spots with the microscope.) Finally, wafer delayering (successive removal of each layer) can uncover the defect or failure site. Identification of the defect often requires a scanning electron microscope and energy dispersive X-ray spectroscopy (EDS) analysis.

AI Wafer Mapping. The first place to start when analyzing failures is to examine the electrical characteristics of the die on the wafer. Harris has a system called AI wafer mapping, which generates a color wafer map. Each die has a color, and each color represents a specific failure bin. Typical failure modes for power MOSFETs include the following:

- IGSS (gate-to-source shorts)
- IDSS (drain-to-source shorts)
- V_{TH} (threshold voltage high or low)
- BVDSS (breakdown voltage high or low)
- RDSON ("on" resistance from drain to source)

An overlay of wafer maps for a lot, or multiple lots, can help pinpoint specific trends or problems. An overlay might reveal that most breakdown voltage failures occur at the edge of the wafer. Figure 11.10 is a greyscale image of a typical wafer map where BIN 8 represents BVDSS failures. Highlighting these failures with *X*s shows that they occur only on the edge of the wafer. (Four are at the right and one is at the bottom.) This knowledge would lead the failure analysis engineer to look at epi thickness or resistivity at the edge of the wafer.

IYM Software. Data management is the hub of the yield management program. It is the tool that connects all of the yield management resources. The IYM software marries the key computer-based data management systems, Workstream, and MEDB into one package. People can access these systems individually, but this time-consuming approach requires separate, unique tools. Mountaintop's in-house IYM software package is available to anyone with access to a personal computer or Sun Workstation.

Statistical Bin Analysis. SBA analyzes the failure modes to identify wafer lots that could have potential reliability problems or low final test yields.

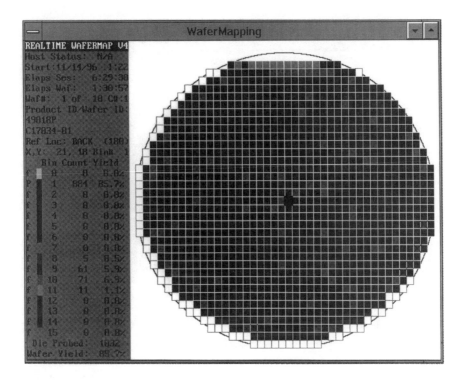

Figure 11.10. AI wafer map of a typical MOSFET showing BIN failures and good die.

Figure 11.11 shows typical failure bins for a MOSFET device; 1237 devices failed, and bin 12 shows the total. Of those, 48 failed because of a test probe continuity error (bin 11). The other 1189 fell into bin 6, 7, 8, 9, or 10. These correspond to threshold voltage, on resistance, breakdown voltage, gate-to-source shorts, and drain-to-source shorts. These data are the averaged data of standard yielding wafer lots for the MOSFET over a one-month period. This composite is, therefore, a good representation of the process capability.

Collection of the data from standard yielding lots is similar to a process capability study for statistical process control. Standards for average and variation come from a process' historical performance. To characterize a process, the engineer collects data while making sure the process is

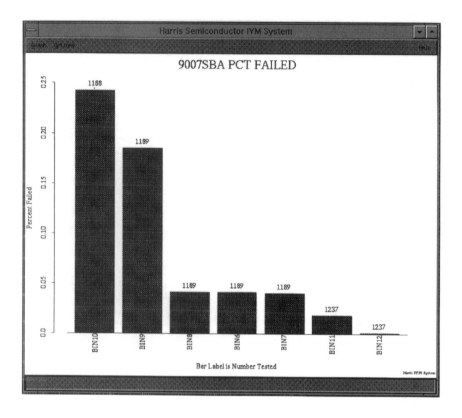

Figure 11.11. Statistical bin analysis of a typical MOSFET device.

in control. These data yield a process average and process standard deviation. Control charts then compare each new sample's average and variation against these standards. A sample is out of control if its average or variation differs too much from the standard. This is evidence of a process shift, which is usually undesirable.

SBA compares each new lot to the historical composite to determine whether the bin levels increase or switch positions on the histogram. Is a particular failure mode becoming more serious? Is the pattern of failure modes changing? Did a process change succeed in reducing a failure mode? SBA is a tool for answering these questions.

Statistical bin analysis is philosophically similar to the multiple attribute control chart. Chapter 12, on statistical methods, discusses a control chart for processes that generate different defects. Instead of tracking the total defects, the chart tracks each defect source to see if it changes.

Process Average Testing. PAT compares parametric data from each lot to a historical composite. The composite comes from good die, and data collection again occurs over a long period. Figure 11.12 is a histogram of threshold voltage for a three-month period on a MOSFET device. PAT is again philosophically similar to statistical process control. It compares each lot or sample against the process' historical performance.

PAT can identify maverick or unusual lots by comparing their parameters against statistically derived control limits. These control limits may range from four to six standard deviations from the average. They are not

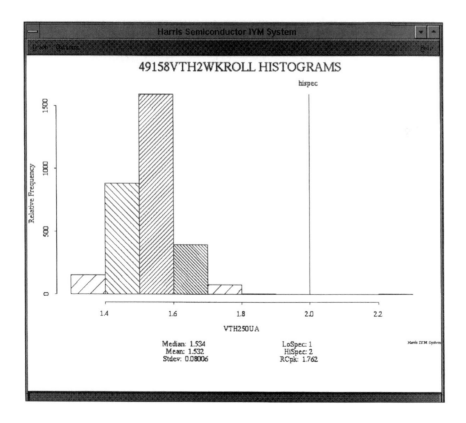

Figure 11.12. PAT histogram of threshold voltage for good MOSFET die.

SPC limits, and they do not check the process for control. Their purpose is to identify statistically unusual product lots. Even if the lot is in specification, violation of these limits implies a problem with the lot. Mountaintop's practice is to have a product engineer disposition such lots.

Liquid Crystal Analysis. Liquid crystal is a material whose transparency changes with temperature. With power MOSFETs, failures like IGSS or IDSS often result from particulate defects or photolithography defects. These defects cause higher-than-normal currents to flow from gate to source or drain to source. This creates a hot spot at the defect site. A thin film of liquid crystal makes it easy to locate the hot spot with a high-power optical microscope. A microprobing station with an optical microscope and 20 mil (0.51 mm) electrical probe tips probes the die and forces current through the failure site. The probe station also stores the defect's coordinates. A typical failure may result from a 0.5-micron particle. With liquid crystal, however, this site will be very visible as a large, circular rainbow pattern.

Chemical Delayering. The probe station computer's memory stores each defect's X-Y coordinates. The failure analyst removes a layer of material and places the wafer back on the probe station. The probe station moves to each defect's position, and delayering will eventually reveal each defect's source. Optical microscopy often reveals a particle or defect in one of the device's layers. From the top down, a typical MOSFET consists of a silicon nitride overcoat, source and gate metal, boron phosphorus silicon glass (BPSG) dielectric, polysilicon gate, gate oxide (SiO_2), and implanted epitaxial (see Figure 11.13). Carbon tetrafluoride plasma, or other fluorine-containing plasmas, will etch silicon nitride. The metal is soluble in hot hydrochloric or phosphoric acid. Hydrofluoric acid etches BPSG, and potassium hydroxide removes polysilicon. A final stain of the silicon with Wright Etch will bring out the implanted regions of the silicon.

Cross-Section Analysis. The failure analyst sometimes needs to know how the implanted dopants diffused into the silicon. How well did an insulator like BPSG separate conductive layers of metal and polysilicon? A mechanical cross-section provides a good look at the silicon below the surface, and answers these questions.

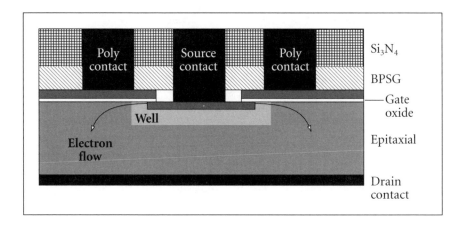

Figure 11.13. MOSFET cross-section.

There are two methods for cross-sectioning. One can simply snap the brittle wafer through the transistor region. The other method is to mount the device on a paddle and lap and polish the transistor on a grinding wheel. Etching with a mixture of HNO_3:HF:acetic acid then highlights each layer. The dopant type (boron, phosphorus, or arsenic) and concentration controls the layer's etch rate. Etching enhances the visibility and contrast of each layer. The scanning electron microscope (SEM) provides a high-power photo of the cross-sectioned device. The SEM also has a secondary electron detector that can help identify the elements on the wafer surface. The technique is EDS.

Time Slide Analysis. Time slide analysis is another important tool for the yield enhancement engineer. It represents a paradigm shift in yield thinking, and it links the fab with probe data. This technique links each manufacturing tool with the wafers' yield at probe. It can help pinpoint tools in the fab that are causing yield losses. Consider the following example.

Probe yield has averaged around 90 percent for the past year. The yield for a particular product, however, takes a sharp drop. The IGSS (gate-to-source shorts) bin is responsible for the sharp yield decrease. How do we know what caused this loss? We can start by asking about the process history.

Did a common tool process all the low-yielding lots? The critical layers for IGSS are gate oxide growth, polysilicon deposition, and metal etch. Probe parametric data show that all of the low-yielding lots went through the same polysilicon deposition tube over a two-day period. Also, these lots failed at probe for IGSS. Further analysis reveals that the tube received improper maintenance before the two days in question. This type of feedback is extremely valuable to the fab and processing engineers.

Figure 11.14 shows how to track the value of a particular parameter, such as IGSS or IEBO,* for specific equipment. Here, a comparative box plot shows IEBO for each lot. The box plot shows the polysilicon deposition tube (Tube N1, Tube Q1, Tube Q2) that processed each lot. The lot in the far left section went through N1. The lots in the middle section went through tube Q1, and most lots went through Q2. The data show no correlation between IEBO and polysilicon tube. In a process that is under control, choice of workstation should not affect the result.

Fortunately, yield busts of this type are rare, but the proper short-loop controls at the fab's critical processes can prevent them. Also remember that if the plant is working at capacity, incidents like these waste irreplaceable capacity. Their true cost is the opportunity cost of making and selling good product. By reducing their frequency, IYM supports the goals of synchronous flow manufacturing. Systematic failure analysis is both a short-loop control and a long-loop control for pinpointing root causes of particular problems in the fab.

Tencor 7600 Particle Monitor

The Tencor 7600 is a particle monitoring system that uses laser technology to locate defects on a wafer. An operator classifies the defects before and after key operations in the fab. A computer database maps and stores the classification and X-Y location of these defects for future reference. Electrical probing identifies die that do not meet electrical specifications. Overlaying the defect map on the probe bin loss wafer map correlates defects with failed die. It associates each defect with a failure type. Failure

---

*Emitter-base current, measured with the collector open, for a bipolar transistor. I refers to emitter-base current. The current between the emitter and base should be minimal, since the base is the transistor's electrical valve. For a MOSFET, this would be gate-source current, where the gate is the electrical valve.

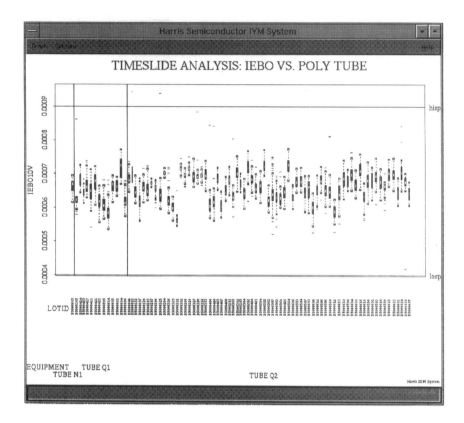

Figure 11.14. IEBO versus polysilicon tube.

analysis of the bad die shows whether the defects were fatal or merely nuisances. Identification of killer defects lets the yield enhancement engineer provide feedback to the process engineer responsible for the tool or process where the defects originated. This feedback can help identify common cause losses that affect all product lines. Remember that correction occurs when we fix special or assignable causes. Improvement requires removal of common causes.

Figure 11.15 shows a defect after gate oxidation. This is the kind of defect the Tencor 7600 can find. This particle later caused a polysilicon-to-source short and an IGSS loss at probe. (The well should separate the conductive polysilicon from the source. The particle provided a conductive

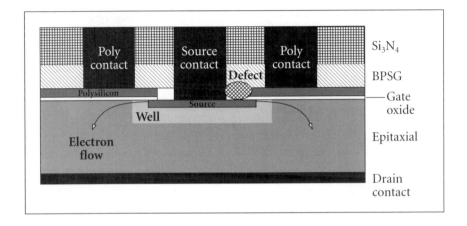

Figure 11.15. Defect after gate oxidation.

bridge.) Analysis of the Tencor defect map and the probe wafer map helps identify this defect as a killer defect.

Test Element Groups

Test element groups (TEGs) have several uses. Two key ones are to provide the engineer with process quality feedback and device reliability feedback. TEGs are diagnostic devices on product wafers, and they are testable during the fabrication process. (Product devices are testable only at wafer probe.) TEGs allow engineers to get electrical parametric data before the wafers reach the end of the line, thus shortening the feedback loop. Typical process quality measurements include the following:

• Sheet resistivity of the implants

• Contact resistance of metal to polysilicon and implants

• Capacitance and breakdown characteristics of gate oxide

• Photoresist mask level-to-level misalignment.

Wafer level reliability measurements include mobile ion contamination, metal stress migration, gate charge trapping, and hot carrier testing. A typical test structure and its cross-section appear in Figure 11.16. This transistor structure provides quality feedback to the etch and implant process engineers.

TEG: HEXPWBSx7
SOURCE-BODY-
WELL-EPI, BACK-
TO-BACK DIODES

NOTE: Contacts are made to poly, source inside well, and epi. Tests on this TEG include:
(1) Vblu_Hex_PWBSx7: Force 1μA of current from source to epi and measure breakdown voltage.
(2) Vbrlu_Hex_PWBS7: Force 1μA of current from epi to source and measure breakdown voltage.
(3) Lk15V_Hex_PWBS7: Force 15 V from epi to source and measure leakage current.

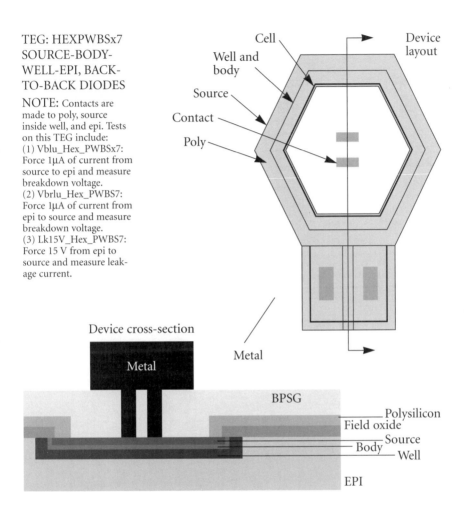

Figure 11.16. Test element group.

Conclusion

Yield management is a discipline that branches all of the organizations at Mountaintop. It incorporates a heavy dependence on teamwork to maintain and improve yields.

CHAPTER 12

Statistical Methods

William A. Levinson

A comprehensive treatment of industrial statistics would require at least an entire book, and this chapter provides some references. This chapter discusses key ideas and principles for using statistical tools effectively. If it gives the reader an awareness of the key issues, it has achieved its purpose. The first part of the chapter discusses some general principles and how Harris Mountaintop teaches these principles to employees.

Many industrial problems do not follow traditional models and assumptions. The second part of the chapter gives readers the benefit of Harris Semiconductor's experience with real-world problems. Again, this chapter cannot provide detailed instructions for handling them, but it provides some references. The goal is to promote awareness of these non-ideal applications.

Key Issues

Harris' experience shows that industrial practitioners must pay close attention to the following issues.

1. Users of statistical techniques must understand *hypothesis testing* and its associated risks. Significance levels are often difficult to understand, and Mountaintop's in-house courses use figures and diagrams to explain the idea.

   —All statistical tests carry a risk (the significance level) of concluding that an effect or problem is present when it isn't.

245

—Statistical tests also carry a risk of not detecting an effect or problem when it is present. The test's power (ability to detect the effect or problem) increases as the effect or problem increases. Increasing the significance level makes the test more powerful, but it also causes more false alarms. "There ain't no such thing as a free lunch (TANSTAAFL)": The only way to increase power without also increasing the false alarm rate is to use a bigger sample.

2. Managers, engineers, technicians, and operators must know the value of design of experiments (DOE). Many industrial experiments are ineffective, or produce misleading results, because of improper design.

3. Traditional statistical process control (SPC) training assumes that manufacturing processes conform to the normal (Gaussian) distribution. Most SPC software programs, and most practitioners, rely on this assumption when they set up their control charts. Real manufacturing processes, however, often do not cooperate with this assumption. Harris Semiconductor has paid close attention to these situations and has adopted methods for handling them.

4. Traditional attribute control charts treat defects and nonconformances as fungible commodities. That is, the charts track total defects or nonconformances. When there is more than one defect or failure type, multiple attribute control charts use the data more effectively.

Hypothesis Testing

Hypothesis testing is the keystone of statistical experimentation and SPC. To use statistical methods effectively, engineers, technicians, managers, and frontline workers must understand this concept.

Harris Mountaintop's statistical training courses emphasize the hypothesis test concept. All statistical techniques—including acceptance sampling, statistical process control, and design of experiments—are hypothesis tests. There are two critical points.

1. Statistical tests separate significant effects from mere luck or random chance.

2. All hypothesis tests have unavoidable, but quantifiable, risks of making the wrong conclusion.

Statistical tests always involve type I (producer's or alpha) and type II (consumer's or beta) risks. The type I risk is the chance of deciding that a significant effect is present when it isn't. (In statistics, *significant* means nonrandom, or not due to chance. A statistically significant effect may not be economically or scientifically significant.) The type II risk is the chance of not detecting a significant effect when one exists. The chapter will explain these shortly.

Common Misunderstandings and Misconceptions
Industrial statisticians often hear questions or statements like these.

• "The last point on the SPC chart was out of control. We'll watch the next run, and see if it comes back inside the control limits."

• "Does a 5 percent significance level mean there is only a 5 percent chance that my results are significant?"

• "The rejection rate was 4 percent, but we think we've improved the process. We ran 50 pieces, and they were all good. If the rejection rate was still 4 percent, we should have gotten two bad ones."

These perceptions all result from lack of knowledge about hypothesis testing. At Harris Mountaintop, we always include hypothesis testing in courses on SPC or experimental design.

Null and Alternative Hypothesis
Every statistical test tests the *null hypothesis* H_0 against the *alternative hypothesis* H_1. *Null* means "nothing," and the null hypothesis is that nothing is present: The process change or treatment makes no difference, or the process is operating properly. The alternate hypothesis is that the process change or treatment has an effect, or something is wrong with the process.

The type I risk is the chance of rejecting the null hypothesis when it is true. The *producer's risk* is the type I risk for an acceptance sampling plan. It is the risk of rejecting a lot that meets the customer's requirements. The type II risk is the chance of accepting the null hypothesis when it is false. The *consumer's risk* is the type II risk for an acceptance sampling plan. It is the chance of passing a lot that does not meet the requirements. If the type I risk is the chance of crying wolf, the type II risk is the chance of not seeing a real wolf. Table 12.1 explains hypothesis testing and risks.

Table 12.1. Statistical hypothesis tests.

	Decision	
State of nature	Reject the null hypothesis	Accept the null hypothesis
The null hypothesis is true. 1. There is no wolf. 2. The defendant is innocent. 3. The lot is good. 4. The process is in control. 5. The treatment has no effect.	Risk = 100α percent, where α is the *significance level* or *producer's risk*. 1. The boy cries "Wolf!" 2. Convict an innocent defendant. 3. Reject a good lot. 4. Call the process out of control when it is in control. 5. Conclude that an experimental treatment has an effect when it doesn't.	There is a $100(1 - \alpha)$ percent chance of accepting H_0 when it is true. 1. The boy doesn't cry "Wolf!" 2. Acquit an innocent defendant. 3. Accept a good lot. 4. Call the process in control when it is. 5. Conclude that a treatment has no effect when it doesn't.
The null hypothesis is false. 1. There is a wolf. 2. The defendant is guilty. 3. The lot is bad. 4. The process is out of control. 5. The treatment has an effect.	The power of the test γ is the chance of detecting the problem or effect. $\gamma = 1 - \beta$ 1. The boy sees the wolf. 2. Convict a guilty defendant. 3. Reject a bad lot. 4. Call the process out of control when it is. 5. Conclude that a treatment has an effect when it does.	The type II risk β is the chance of not detecting the problem or effect. 1. The boy doesn't see the wolf. 2. Acquit a guilty defendant. 3. Accept a bad lot. 4. Call the process in control when it isn't. 5. Conclude that a treatment has no effect when it does.

Significance Level in Hypothesis Testing

Experiments return statistics such as z (the standard normal deviate), t, F, or χ^2 (chi square). The statistic depends on the experiment, but they all measure the evidence against the null hypothesis. We reject the null hypothesis for large values of $|z|$, $|t|$, F, or χ^2, and accept it for small ones. Statistical tables specify what is "large" for the particular experiment or test. It depends on the available information and the selected significance level.

Figure 12.1 shows a graphical explanation of significance level. Figure 12.1 is for an experiment that returns χ^2 with six degrees of freedom. The 95th percentile for a chi square distribution with six degrees of freedom is 12.59. If the null hypothesis is true, there is a 95 percent chance that χ^2 will be less than 12.59.

Don't worry about Equation 12.1; it is only for reference. Figure 12.1 is more important. If the null hypothesis is true, 95 out of 100 experiments will generate a χ^2 in the unshaded part of the distribution. If χ^2 is greater than 12.59, the experimenter can be 95 percent sure there is a

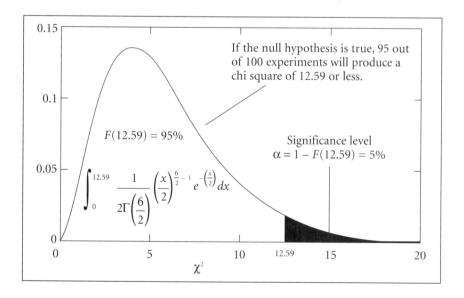

Figure 12.1. Chi square distribution with six degrees of freedom.

genuine effect or problem. There is a 5 percent chance of getting lucky (or unlucky) and getting a χ^2 in the shaded region.

For χ^2 with υ degrees of freedom,

$$F(\chi_\upsilon^2) = \int_0^{\chi_\upsilon^2} \frac{1}{2\Gamma\left(\frac{\upsilon}{2}\right)} \left(\frac{x}{2}\right)^{\frac{\upsilon}{2}-1} e^{-\left(\frac{x}{2}\right)} dx \text{ and } \alpha = 1 - F(\chi_{\upsilon;\alpha}^2) \quad \textbf{(Eq. 12.1)}$$

$\chi_{\upsilon;\alpha}^2$ is the $100(1 - \alpha)$ percentile of the chi square distribution with υ degrees of freedom.

Pictures and Diagrams for Statistical Training

Diagrams like Figure 12.1 are very useful for explaining significance levels and testing risks. Harris uses such diagrams extensively in its statistical training programs. This one is from MathCAD 6.0, but earlier versions of MathCAD can produce the same figure. One can then copy and paste it into Microsoft Word (and probably Lotus Word Pro and Corel WordPerfect) to make handouts. Microsoft Draw or a similar drawing editor can add explanatory notes to the figure. Microsoft PowerPoint (and probably Lotus Freelance and Corel Presentations) can accept the figures for classroom instruction. Harris Semiconductor used PowerPoint to make the transparency set that goes with Levinson and Tumbelty's (1997) *SPC Essentials and Productivity Improvement: A Manufacturing Approach.* CorelDraw 3.0 (or higher) can convert the MathCAD figure into a .GIF (graphics interchange format) file for display on the Internet. The Mountaintop plant has also used Manugistics' StatGraphics (Windows version) and Microsoft Excel to draw figures for training materials.

Type II Risk and Power in Hypothesis Testing

The type I risk is the chance of seeing wolves when they aren't there. The type II risk is the chance of not seeing the wolf. We might expect this to depend on how close the wolf is to the sheep. As the wolf gets closer, the boy is more likely to see it. It's the same in hypothesis testing: A test is more likely to detect a big effect (or problem) than a small one. *A test's power is its ability to detect a process problem or an effect.* It depends on the effect's magnitude and the sample size.

The example of a court verdict helps explain the idea. An acquittal does not prove the defendant innocent. It means there is not enough evidence for a conviction beyond a reasonable doubt. In statistics, the "reasonable doubt" is the type I risk. Accepting the null hypothesis means we cannot be $100(1 - \alpha)$ percent sure there is a problem or an effect. The problem or effect may be there, but we cannot detect it.

In SPC, it is not acceptable to "wait and see if the process comes back into control." The presence of a point inside the control limits does not prove that the process is in control. A point outside the limits, however, is strong evidence that it isn't. Therefore, manufacturing personnel must respond to each out-of-control signal.

Consider the example of a 4 percent rejection rate and a sample of 50 pieces. We expect to get two bad ones. What, however, is the chance of getting none? There is a 0.96 chance that each piece will be good. There is a $0.96^{50} = 0.130$ (13.0%) chance that all 50 will be good. If we think we've improved the process, run 50 pieces, and all are good, there is a 13.0 percent chance that it's only luck. Five percent is a common significance level in hypothesis testing. With a sample of 50, we cannot be more than 87 percent sure that the improvement worked. The experiment lacks the power to detect the process improvement because the sample is too small.

Design of Experiments

Design of experiments (DOE) is a systematic approach to scientific or industrial experiments. Harris Semiconductor encourages employees to use DOE to improve quality and solve problems. Mountaintop trains engineers and technicians in the standard techniques and principles.

Design of Experiments: Benefits

In the mid-1980s, DuPont Corporation developed a diagnostic test for AIDS.

> To meet the Food and Drug Administration's requirements for specificity and sensitivity, researchers needed to sort out the effects and interactions of 19 separate variables. Through a four-week planned experiment, they succeeded

in producing a test that was better able than competitive products to detect the AIDS virus in the early stages of infection. (Cusimano 1996)

DuPont used an experimental design to overcome a tough technical challenge and beat its competitors. The reference does not say what technique DuPont used, but it was probably a fractional factorial design. This is the most logical approach for handling a system with 19 independent variables and interactions. In contrast, the bankrupt "one variable at a time" (OVAT) approach wouldn't have had a chance of solving the problem. Unfortunately, many people still use this approach (see Figure 12.2). Cusimano (1996) continues,

Academicians estimate that from one in 20 to one in 100 experimenters who could benefit from DOE actually use it. Instead, researchers often use the classic "one factor at a time" technique. Using this approach, researchers alter the setting of one variable, hold the others constant, and then measure the resulting responses.

Interactions

Figure 12.3 shows examples of interactions in the model $y = f(x_1,x_2)$. The one-factor-at-a-time approach will work if there are no interactions, and the model is $y = \beta_0 + \beta_1 x_1 + \beta_2 x_2$ where the βs are coefficients. (β_0 is the intercept.) The coefficient for x_1 is independent of x_2, and vice versa. If the experimenter holds x_2 constant while adjusting x_1, he or she can estimate β_0 and β_1. The same process, with x_1 constant, yields an estimate of β_2.

OVAT fails, however, when there is an interaction. The model is then $y = \beta_0 + \beta_1 x_1 + \beta_2 x_2 + \beta_{12} x_1 x_2$ or $y = \beta_0 + \beta_1 x_1 + (\beta_2 + \beta_{12} x_1)x_2$ or $y = \beta_0 + (\beta_1 + \beta_{12} x_2)x_1 + \beta_2 x_2$. That is, the slope coefficient for x_2 is a function of x_1, and vice versa. If the interaction is negative, and the experimenter holds x_2 constant at a low level, the coefficient for x_1 will be positive. If, however, the experimenter keeps x_2 high, the coefficient for x_1 will be negative. The slope coefficient for x_1 varies with x_2, and vice versa. OVAT, therefore, can waste time and produce misleading results.

Figure 12.2. OVAT experimentation.

Probabilistic Versus Deterministic Systems

Cusimano (1996) says the underlying problem is that engineers and scientists do not learn design of experiments in college. In the physical sciences and engineering, problems usually have deterministic answers: $y = f(X)$, where y is the response variable and X is a vector of design, or independent,

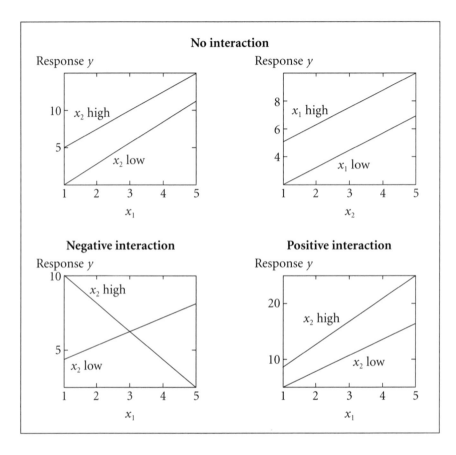

Figure 12.3. Interactions.

variables. Manufacturing systems are, however, usually probabilistic: $y = f(X) + \varepsilon$, where ε is a random number. Table 12.2 shows the difference between probabilistic and deterministic systems.

This comparison leads to an obvious question. "We design the system according to $y = f(X)$, but the real world is $y = f(X) + \varepsilon$. Does this mean everything we learned about engineering design and physical systems is wrong?" Although $y = f(X) + \varepsilon$, $y = f(X)$ is the best bet. If we perform the operation repeatedly, the average result will be $f(X)$. There is a specific term for the average result. It is $E(X)$, or "the expected result for X."

Table 12.2. Determinate systems versus probabilistic systems.

	Determinate System	Probabilistic System
Typical applications	Engineering design, physical and chemical equations	Manufacturing systems, experiments
Response or dependent variable	$y = f(X)$	$y = f(X) + \varepsilon$, where ε is random
Knowledge of y	100 percent sure it's $f(X)$	95 percent* sure it's in the range $[a,b]$, which depends on $f(X)$ and the uncertainty in the system.

*95 percent is an example. We can compute other confidence intervals as well.

A probabilistic system implies a repetitive operation or test. Infinite trials under the same conditions will yield an average of $E(X)$. Ninety-five percent of the results (y) will be in a range $[a,b]$, where a and b depend on the probabilistic system. If the system follows a normal or Gaussian distribution, a and b depend on its mean (μ) and standard deviation (σ). Eighty percent, 90 percent, 99 percent, and other population percentiles also can be computed. If a and b are the specification limits, we can compute the fraction of the product that will conform to them (see Eq. 12.2).

$$\text{Fraction within } [a,b] \;=\; \int_a^b \frac{1}{\sqrt{2\pi}\sigma} \exp\!\left(-\frac{1}{2}\left(\frac{x-\mu}{\sigma}\right)^2\right) \qquad \textbf{(Eq. 12.2)}$$

Multiple Trials Versus Single Trials

Repetition is how casinos—and insurance companies—make money. Suppose a roulette wheel has 38 slots, and the game offers 35 to 1 odds. (There are 36 numbers, plus 0 and 00.) Someone bets $100 on a number and spins the wheel. There is a 1 in 38 chance that the ball will land on the number, and the casino must pay $3500. There are 37 chances that the casino will win the $100. The decision tree in Figure 12.4 shows that, on average, the casino will win $5.26. Would you offer these odds on a one-time event? Most people wouldn't, because the small chance of losing $3500 outweighs the benefits of winning $100.

In Clint Eastwood's famous movie *Dirty Harry*,* Harry Callahan pointed a .44 Magnum pistol at a robber. The villain thought the detective had used all his ammunition and was thinking of going for his own gun. The detective reminded him that a .44 Magnum "can blow your head clean off," and asked, "Do you feel *lucky*?" Since this was a one-time trial, and not a repetitive process, the bad guy surrendered. The .44 Magnum

---

*Warner Brothers, 1971. Directed and produced by Don Siegel. Screenplay by Harry Julian Fink.

The famous lawyer Gerry Spence (1995), author of *How to Argue and Win Every Time*, says that stories are a powerful communication medium. Mountaintop's training programs use stories or well-known news events to teach statistical methods. For the benefit of readers outside the United States, Clint Eastwood is a famous actor in American Westerns and detective stories. Many American readers remember his line, "Do you feel *lucky*?" from the movie *Dirty Harry*. When we associate a statistical concept with a well-known story, people are more likely to remember it.

Mountaintop's SPC training uses the "three strikes" laws (laws that mandate life sentences for offenders who commit three or more violent crimes) to demonstrate the Pareto Principle: A few causes (criminals) are responsible for most of the rework and scrap (crime).

Mountaintop used the famous O.J. Simpson trial to emphasize the ISO 9000 requirement for gage calibration. The Los Angeles police left Simpson's Ford Bronco unattended on the police evidence lot, and someone broke into it. This violated the chain-of-custody requirement for legal evidence, and may have lost the trial for the prosecution. Similarly, manufacturing personnel cannot rely on data from a gage that is out of calibration. They cannot trust any data gathered since the gage's last calibration date. Any product shipments since the gage's last calibration are suspicious, since they may have been checked with a bad gage.

Muskets and rifles demonstrate high and low variation processes; the musket is an example of a noncapable tool. Stories are useful teaching devices, and they become especially important when teaching nonmathematical audiences. If the instructor throws a lot of mathematical formulas at manufacturing operators, he or she will lose most of them. The audience will, however, understand the difference between a musket's shot pattern and a rifle's. If the idea is to hit the target (specification), the audience will realize that a rifle (low-variation process) is better.

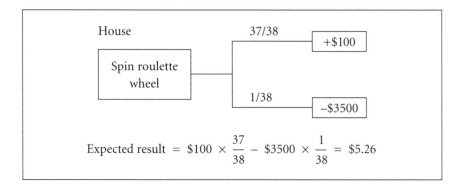

Figure 12.4. Decision tree: Probabilistic situation.

was probably empty, but maybe he had miscounted the shots. He probably imagined himself before the Pearly Gates (or the other place) saying, "On average, I made the right choice." This was actually an uncertain system, not a probabilistic one (see Figure 12.5). In an uncertain system, the odds are unknown. This is why insurance companies hire actuaries: If we don't know the odds, we should not play the game.

In Figure 12.5, "Head blown clean off" is the same as, "Lose 35-to-1 roulette wager" or "Pay massive insurance claim." This is why casinos don't want to gamble on single events, and why insurers don't like to cover single events. Gambling on single events raises the question, "Do you feel lucky?" The roulette wheel becomes Russian roulette, with a .44 Magnum round in one of the 38 chambers. This is all right if the wheel's proprietor can play forever and buy a new head with every 35 successes. It is a very poor one-time proposition.

Casinos do, of course, play the same game tens of thousands of times. They usually win $100, and occasionally lose $3500. On average, the casino wins $5.26 on this bet. Here, $E(X)$ is $5.26, although outcomes range from +$100 to –$3500. Insurance companies that cover large populations have to pay some claims, but on average they make money. Insurance companies would always rather cover large populations than small ones. For example, it is less risky to insure 100,000 lives than five communications satellites. Actuaries can be 99 percent confident that

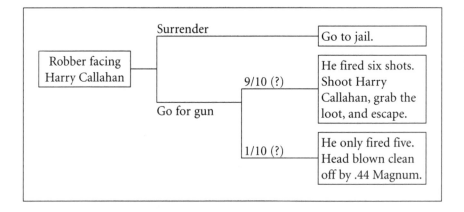

Figure 12.5. Decision tree: Uncertain situation, one-time event.

between *a* and *b* people will die in any year, but no one knows the exact chance that the rockets will blow up. ("Do you feel lucky?")

Manufacturing operations that make thousands of identical parts are repetitive, and they are amenable to statistical methods. While $y = f(X) + \varepsilon$, where ε is a random number, $f(X)$ is the best bet. Also, when we know how ε behaves, we can calculate how often we will win (be in specification). A designed experiment characterizes the function f and sorts it out from ε, the random noise.

Why Don't More People Use Statistical Methods?

Engineers are usually aware of random variation, which is why they design safety factors into structures. Both the strength of the material, and the stresses to which it is subject, can vary. If an unusually low strength runs into an unusually high stress, the material will fail. It is of little comfort to someone whose building or bridge collapses that an average girder would have supported an average load.

Even the safety factor approach must rely on experience if statistical data are unavailable. "The failure probability may vary from a low to an intolerably high value for the same safety factor" (Kapur and Lamberson 1977, 73–75). The underlying statistical models are, however, graduate-level material. Figure 12.6 shows the situation for normally distributed* material strengths and stresses. The material's strength is, on average, 10 units greater than the stress. Should the designer be comfortable with this situation?

Typical DOE courses are graduate-level statistics courses, with statistics prerequisites. Therefore, only people who are majoring in applied statistics take them. The result is that engineers and scientists often lack education in industrial statistics. Cusimano (1996) quotes Gerald Hahn, manager of General Electric's corporate research and development statistics program in Schenectady, NY.

> *The concepts are somewhat foreign to what most engineers learn in school. Varying one factor at a time is ingrained, so looking at a number of dimensions simultaneously is contrary to what they know . . . We offer DOE courses, but with prerequisites, so only statisticians end up taking them.*

This is why, as Cusimano says, most people who can benefit from DOE do not use it. If this section creates an awareness of DOE, it has accomplished its purpose.

Design of Experiments: Overview

Experiments have two basic goals.

1. Identify or quantify a treatment's effect. Answer questions like, "Is the new process better than the existing process? If so, by how much?" Quantify (fit) and verify the function $y = f(X)$ and show how the design variables affect the response. Which design variables are important?

2. Some design variables may not affect the result. Distinguish systematic effects from random variation or noise. "Did the experiment work, or did we just get lucky?"

---

*Primarily for simplicity. The distribution could easily be nonnormal. The extreme value model assumes that the material fails at its weakest point.

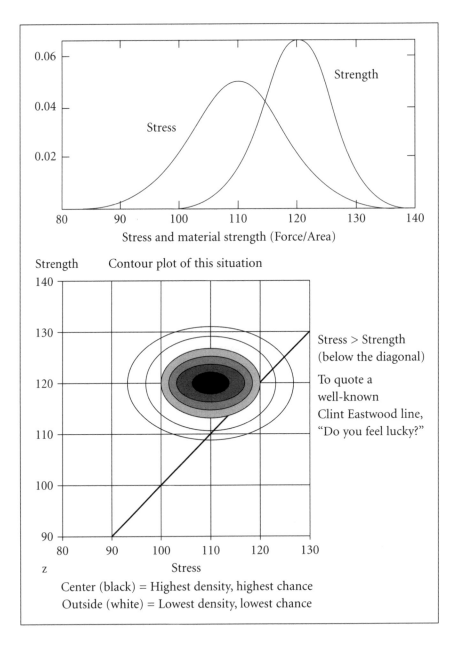

Figure 12.6. Probabilistic nature of material strengths and stresses.

DOE can perform the following tasks.

1. Experiments can characterize physical or chemical process models. In an etching process, acid concentration and temperature affect the rate. For a spin coating process, spin speed, spin time, and solution viscosity affect the coating thickness.

2. Experiments can determine which process factors affect quality, reliability, and yield.

3. Experiments can guide adjustments of the process factors to improve quality, reliability, and yield. Response surface methodology (RSM) or evolutionary optimization (EVOP) is an important method for systematically improving a process.

Design and Response Variables. An experiment always involves one or more *design variables* and a *response variable*. Design variables are factors that the experimenter can manipulate, or at least measure. The response variable measures the outcome, or result. An experiment for p parameters looks like this in equation form: $y = f(X) = f(x_1, x_2, \ldots, x_p)$ where y is the response variable (outcome) and x_j is the jth design variable. The outcome or response y is a function of the design variables.

The Linear Model. Most experimental designs use a linear model. Linear regression, analysis of variance (ANOVA), factorial designs, Latin squares, and Youden squares are all linear models. For two design variables, the linear model is

$$y = \beta_0 + \beta_{11}x_1 + \beta_{21}x_2 + \beta_{1121}x_1x_2 + \beta_{12}x_1^2 + \beta_{22}x_2^2 + \ldots$$

where coefficient β_{ik} is for x_i to the kth power.

We can break down the parts of the equation as follows:

$$\overset{\text{Interaction}}{y = \beta_0 + \underset{\text{Linear terms}}{\beta_{11}x_1 + \beta_{21}x_2} + \beta_{1121}x_1x_2 + \underset{\text{Quadratic terms}}{\beta_{12}x_1^2 + \beta_{22}x_2^2} + \ldots}$$

This model is *always* valid over short ranges of the design variables. It is, however, dangerous to extrapolate outside the ranges of the design variables. The Taylor series proves that we can express any unknown function

$f(X)$ as a linear polynomial, *over short ranges of the design variable.*[*] The Taylor series extends to functions $f(x_1, x_2, \ldots, x_p)$ for p design variables, so these also are expressible as linear models (with interaction terms). For practical applications, terms rarely go beyond quadratic, at least in the exploratory phase.

The t test can reject terms that are not significant, thereby simplifying the model. The process of selecting terms that belong in the model is *model building.* References on regression and design of experiments provide details.

Attributes and Variables. The variables can be quantitative (real numbers) or qualitative (present/absent, good/bad, material A, B, or C, and so on). *If given a choice, we always prefer quantitative numbers over qualitative results.* Attributes are qualitative: good or bad, yes or no, one or zero. Defect counts, which use a whole number scale, are also attributes. Variables are quantitative measurements: microns, mils, °C, ohms, amperes, pounds per square inch, and so on. They are continuous scale, or real number, measurements.

Optimized statistical methods are available for treating attribute data. One quantitative measurement, however, is often worth 10 attribute data, or more. Figure 12.8 shows 97.5 percent upper confidence intervals for the nonconforming fraction when all n pieces are good. The practice of testing n pieces and passing the lot if there are no failures is a "zero acceptance number sampling plan." While it sounds strict, it cannot ensure high quality levels. If 500 pieces are good, the manufacturing personnel can be 97.5 percent sure only that the nonconformance rate is less than 0.736 percent. (There is a $0.99264^{500} = 0.0249$ percent chance of getting no bad pieces if the nonconformance rate is 0.736 percent.)

Suppose the experimenter wants to know whether a process change has improved the yield (reduced rework and scrap). The goal is to be 95 percent sure that a lower rejection rate is because of the change, and not mere luck. The factory will run n pieces and conclude that the change

---

[*]For a function in one variable, when x is near a,

$$f(x) = \sum_{k=0}^{\infty} \frac{1}{k!} f^{(k)}(a)(x-a)^k = f(a) + \frac{df}{dx}\bigg]_a (x-a) + \frac{1}{2}\frac{d^2f}{dx^2}\bigg]_a (x-a)^2 + \ldots$$

Only sampling plans that use numerical measurements (variables) can ensure extremely high outgoing quality.

Sampling plans with zero acceptance numbers (inspect *n* pieces, fail the lot if any are bad) are strict, and they ensure high quality levels.

The Gordian Knot:
"Can't be done"
"Always did it that way"
"It will never work"

Beth Hollock, 1997

Figure 12.7. Zero acceptance number sampling and quality.

worked if none are bad. Table 12.3 shows how many pieces the factory must make to do this.

Suppose the rejection rate is one part per thousand, which is not very good by modern standards. To check whether a process improvement reduced this, the factory must run 2995 pieces. If the change was not effective, there is only a $100(1 - 0.001)^{2995} = 4.996 \approx 5$ percent chance of getting lucky and having no bad ones. If all 2995 are good, manufacturing personnel can be 95 percent sure that the new nonconformance rate is less

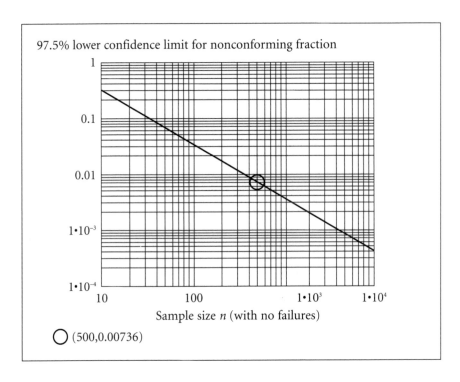

97.5% lower confidence limit for nonconforming fraction

Sample size n (with no failures)

○ (500,0.00736)

Figure 12.8. 97.5 percent lower confidence interval for the nonconforming fraction.

Table 12.3. Sample needed for 95 percent confidence that the rejection rate is lower.

Original nonconformance rate	n
10%	29
1%	299
0.1% or 1 ppt	2995
0.01% or 100 ppm	29,956
0.001% or 10 ppm	299,574
1 ppm	2,995,733

than 1 ppt. However, a process characterization study with as few as 30 numerical measurements can do the same job.* One hundred measurements would be preferable, but it's still 100 pieces versus 2995.

Meanwhile, suppose the process change was successful and reduced the nonconformance rate to 0.5 per thousand. We still expect 1.5 bad pieces out of 2995. We'd probably get one or two bad ones, and conclude that the change didn't work. Not only do we need a huge sample, but the sample might not detect a significant change. Mountaintop's training programs always stress the superiority of variables over attributes.

At the lower nonconformance rates, which are realistic when people talk about six-sigma processes, using attributes would be futile. The necessary sample could easily exceed the item's lifetime production schedule!

Attribute data are the least useful data, but they are sometimes the only available data. Some processes generate only good/bad data or defect counts. Harris has paid close attention to methods, such as the multiple attribute control chart, for getting the most out of attributes. When there is a choice, however, a measurement is always better than an attribute. In summary, use attribute data only when nothing better is available.

Statistical Process Control

Feedback Process Control

SPC is a form of feedback process control. Automatic feedback process control is common in the chemical process industries. The goal of SPC is to allow timely corrections for undesirable process changes. Figure 12.9 (Levinson and Tumbelty 1997, 94) shows a feedback control loop.

The factory must, however, avoid adjusting the process when no change has occurred. Overadjustment, or tampering, means reacting to random variation by adjusting the process. Such adjustments make the variation worse. Mountaintop's training materials compare overadjustment

---

*If the data follow a normal distribution, we can get a lower confidence limit for the process capability index. This can provide 95 percent assurance that the rejection rate does not exceed a certain level. Thirty is usually considered the minimum number of samples for a process characterization (capability study). Technically, fewer than 30 samples could show that an improvement had worked.

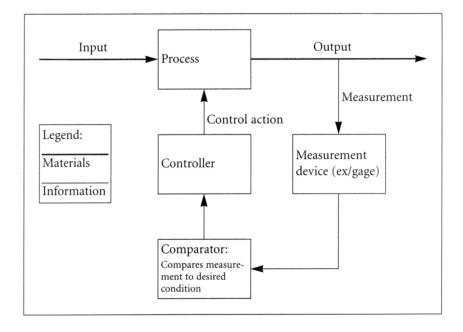

Figure 12.9. Feedback process control loop.

to prescribing medicine for a patient who isn't sick. At best, it is a waste of money, and it can make the patient sick. Tampering can give the process an iatrogenic disease. (*Iatros* is Greek for "physician," and *genic* means "caused by.") The famous funnel experiment, which adjusts a process for random variation, demonstrates the idea. The funnel experiment involves placing a target under a funnel, dropping balls or marbles through the funnel, and recording where they land. Adjustment after every ball produces a pattern similar to Figure 12.10. In Figure 12.10 (Levinson and Tumbelty 1997, 87), if the target center is (0,0) and the shot hits at (x,y), the simulation adjusts the aiming point $(-x,-y)$.

Variation and Accuracy

Variation and accuracy are vital concepts in statistical process control. Harris Semiconductor uses figures liberally to teach them to manufacturing personnel. The instructional materials begin by treating the specification

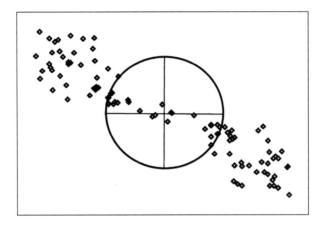

Figure 12.10. Overadjustment or tampering.

as a target and the manufacturing process as a musket or rifle. A musket is a high-variation tool, while a rifle is precise.

The following story shows why a high-volume, high-variation tool might be acceptable. The variation doesn't matter if the target, or specification, is wide. The English Tower musket ("Brown Bess") of the eighteenth century relied on a wide specification. The musket's loose-fitting ball made it easy to load, and a soldier could fire five or six shots a minute. Since the ball fit loosely in the unrifled barrel, aiming was almost futile. An English regiment could, however, discharge 10,000 shots a minute in the enemy's general direction. Since European troops fought in shoulder-to-shoulder formations, they were big targets. Figure 12.11 (Levinson and Tumbelty 1997, 81) uses the French as the victims, since that's usually who the English were fighting in the eighteenth century. The shot pattern comes from a MathCAD simulation and is superimposed on the target's center.

In reality, not everyone would shoot at one opponent. The demonstration shows the chance of each shot hitting its intended target. Since the opponents are in a shoulder-to-shoulder formation, a shot that misses one may hit another. This is like selling off-grade or downgraded product to a customer who will accept it!

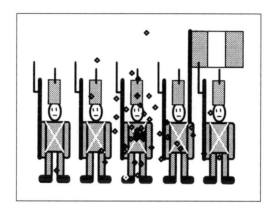

Figure 12.11. High-variation tool versus wide specification.

Most Americans are familiar with the story of the American Revolution. The Revolutionaries did not fight in the open or in shoulder-to-shoulder formations. They knew the dense North American forests and used them to an advantage. The English did not have a big formation of brightly dressed Europeans to shoot at. Instead, they caught glimpses of men in dull jackets and coonskin caps who were hiding behind trees. The specification limits were suddenly much tighter—a situation that many manufacturing companies are experiencing today. Figure 12.12 (Levinson and Tumbelty 1997, 82) shows the situation.

The students in SPC training courses first learn that the chance of hitting the target (specification) depends on the tool's variation and accuracy. Then they are ready to learn about control charts. Figure 12.13 (Levinson and Tumbelty 1997, 116) shows the control charts for an in-control process with a target and a histogram. The target shows a shot pattern from a rifle (low-variation tool) whose aiming point is the target center.

In Figure 12.14 (Levinson and Tumbelty 1997, 117), a musket (high-variation tool) has replaced the rifle. The instructor emphasizes that the manufacturing worker can see only the control charts. He or she must

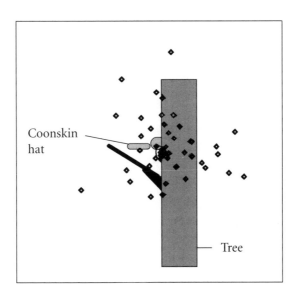

Figure 12.12. High-variation process versus tight specification.

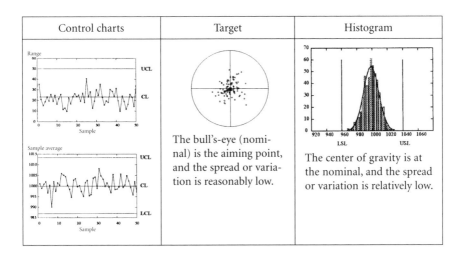

Figure 12.13. Control charts for in-control process.

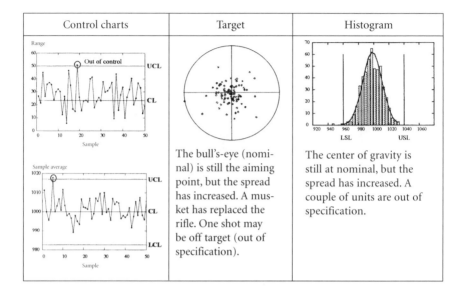

Control charts	Target	Histogram
	The bull's-eye (nominal) is still the aiming point, but the spread has increased. A musket has replaced the rifle. One shot may be off target (out of specification).	The center of gravity is still at nominal, but the spread has increased. A couple of units are out of specification.

Figure 12.14. Control charts after increase in process variation.

infer the process' condition from the charts. When the range (or sample standard deviation) chart is out of control,* the user infers that the target and the histogram appear as shown.

In Figure 12.15 (Levinson and Tumbelty 1997, 118), the process average has shifted. The manufacturing worker infers this from the $\bar{x}$ (sample average) chart.

Requirements for Successful SPC

Table 12.4 shows four requirements for successful SPC (Messina 1987, 1–2).

---

*There is an out-of-control point on the chart for averages because of the excessive process variation. This chart's control limits rely on the process' historical variation ($\mu \pm 3\sigma$ or empirical equivalent), and this variation (σ) has increased. The chart's users would have investigated the fifth sample and found no assignable cause.

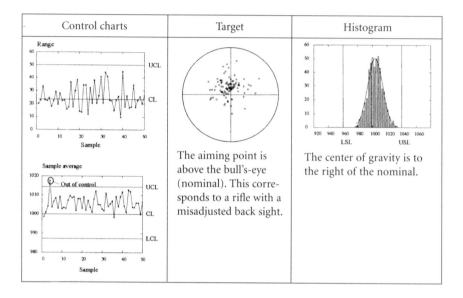

Control charts	Target	Histogram
	The aiming point is above the bull's-eye (nominal). This corresponds to a rifle with a misadjusted back sight.	The center of gravity is to the right of the nominal.

Figure 12.15. Control charts after increase in process mean.

Multiple Attribute Control Charts

We previously showed that attributes are the least useful data. Sometimes, however, they are the only data available. Traditional attribute control charts include the following:

1. *np* chart (number nonconforming)

2. *p* chart (nonconforming proportion)

3. *c* chart (defects)

4. *u* chart (defect density, or defects per piece)

These charts all treat defects as fungible commodities: A nonconformance is a nonconformance, and a defect is a defect.

In practice, however, nonconformances and defects come from different sources. This is why people use tally sheets and make Pareto charts! The traditional charts lose much of the information that is present in tally sheets and Pareto charts. The multiple attribute control chart recovers this information and uses it intelligently. A multiple attribute control chart is

Table 12.4. Requirements for successful SPC.

1. Data integrity	Data (measurements) must be accurate. Instruments must be accurate, and production workers must record measurements accurately. Calibration is an ISO 9000 requirement, and Mountaintop has a systematic program for calibrating production gages. Gages must be capable, and Harris' Mountaintop plant requires periodic gage studies. Reproducibility and repeatability (R&R) studies measure the gage's variation. Capability and accuracy are different. An accurate gage will, on average, return the standard's correct measurement. If there is a lot of variation in the measurements, however, the gage is not capable.
2. Data traceability	Measurements must be traceable to the processes, tools, and materials that produced them. If there are, for example, four workstations, we would keep a separate control chart for each. We would not mingle measurements from different stations on one chart. If the chart shows an out-of-control condition, we want to know which station was responsible.
3. Identify critical process parameters	It is a mistake to keep control charts for every measurement and every operation. No one can pay attention to all the charts, which creates the danger that no one will pay attention to the important ones. When the charts do little other than cover the walls, Hradesky (1988, 183) calls them *wallpaper*. Instead, we must identify the process steps that have significant effects on product quality. Harris Semiconductor calls these operations *critical nodes.*
4. Real-time capability	Feedback must be prompt enough to allow timely process adjustments. Where possible, gages should be at the workstations to provide immediate feedback. If not, the measurement step should be as close as possible to the production step.

simply a tally sheet with control limits. This has several advantages over the traditional charts.

1. The multiple attribute chart is very easy to use. The operator records the data as if on a tally sheet. He or she simply compares the defect or nonconformance count for each trouble source against its control limit.

   —The situation is more complex if the sample sizes vary. If the operator can type the counts into a spreadsheet, however, the spreadsheet can handle the calculations.

2. The chart is more powerful than a traditional chart with the same overall false alarm risk. Since the multiple attribute chart examines each trouble source independently, there is nowhere to hide. The traditional chart lumps all the causes together, so an out-of-control situation can hide for a long time.

   —Lumping all the problem sources together is a lot like plotting measurements from different workstations on the same SPC chart. The key idea is data traceability. The thought of plotting measurements from five drill presses on one $\bar{x}-R$ or $\bar{x}-s$ chart would appall most manufacturing professionals. It is, however, generally accepted practice to pool nonconformance or defect counts on one attribute chart!

Levinson and Tumbelty (1997, chapter 4) shows how to set up and use multiple attribute charts; see also Cooper and Demos (1991) and Levinson (1994a). The key concepts to remember are that: (1) it's a simple tally sheet with control limits and (2) it maintains traceability to each problem source.

Real-World Statistics

Most books rely on the assumption that manufacturing data follow the normal distribution. The normal distribution is often a good model, but it is not universal. Real manufacturing processes are often uncooperative (see Figure 12.16). Most books show how to test the normality assumption, but not how to handle processes that don't conform.

Data from manufacturing processes come from normal (Gaussian, bell curve) populations. We can rely on this assumption to predict process yields, calculate control limits, and so on.

Data from real manufacturing processes often follow nonnormal distributions, or originate from nested variation sources. Don't assume anything, and be aware of these nonideal situations.

The Gordian Knot:
"Can't be done"
"Always did it that
    way"
"It will never work"

Beth Hollock, 1997

Figure 12.16. Normal distribution assumption.

Batch Processes and Nested Variation

If the process is in control, 99.73 percent of the sample averages will be within the 3σ Shewhart control limits. Figure 12.17 (Levinson and Tumbelty 1997, 161) charts data from a process that is in control. Six of the 50 points are outside the control limits.

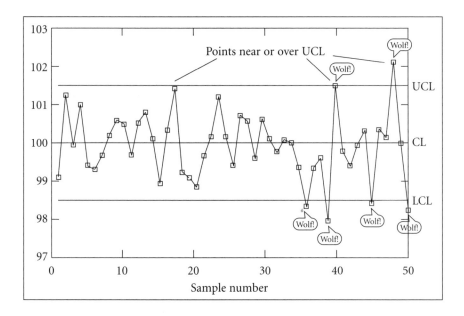

Figure 12.17. Chart for process average, nested variation sources (incorrect).

A chart such as Figure 12.17 can easily convince manufacturing personnel that SPC is useless, or worse. They would have investigated the out-of-control signals and found nothing. The Shewhart control chart promises a 0.27 percent false alarm rate: The boy should cry "Wolf!" two or three times for every thousand samples. In Figure 12.17, the boy has cried "Wolf!" six times out of 50 samples. In Aesop's story, the boy lost his credibility by crying "Wolf!" too many times. Charts such as Figure 12.17 will have the same effect on a factory's SPC program.

Figure 12.18 (Levinson and Tumbelty 1997, 159) shows the source of the control limits for Figure 12.17. Manufacturing workers measured two pieces out of each batch of 12. They used the average range, or average standard deviation, of the two piece samples to estimate the process' variation. The Shewhart control limits are then $\bar{\bar{x}} \pm 3 \dfrac{\hat{\sigma}}{\sqrt{n}}$ where $\bar{\bar{x}}$ is the average of all the data and σ an estimate of the standard deviation. The limits are for the average of a sample of n pieces.

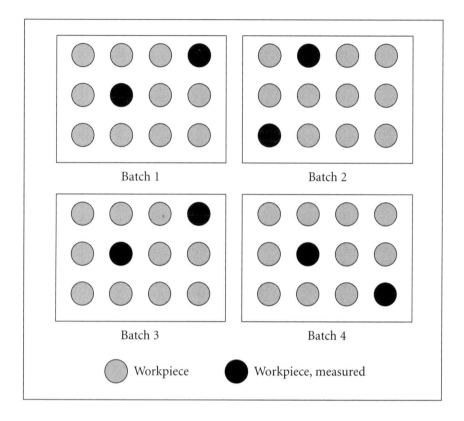

Figure 12.18. Batch process.

Here's what's wrong. A rational subgroup represents a homogeneous set of process conditions. A batch is not a homogeneous representation of the process; there is unavoidable variation between batches. Let x be the measurement for an individual piece. $x \sim N(\mu, \sigma^2)$ is shorthand for "x comes from a normal distribution with mean μ and variance σ^2." $\mu_{batch} \sim N(\mu_{process}, \sigma^2_{between\ batch})$ and then $x \sim N(\mu_{batch}, \sigma^2_{within\ batch})$. If $\bar{x}$ is the average for n pieces, $\bar{x} \sim N\left(\mu_{process}, \sigma^2_{between\ batch} + \frac{1}{n}\sigma^2_{within\ batch}\right)$. The control limits should be $\mu_{process} \pm 3\sqrt{\sigma^2_{between\ batch} + \frac{1}{n}\sigma^2_{within\ batch}}$. Treating n pieces per batch as a "subgroup of n" yields $\mu_{process} \pm 3\sqrt{\frac{1}{n}\sigma^2_{within\ batch}}$, and that's why the control limits are too tight! The problem becomes worse as n becomes bigger.

The problem extends to the chart for process variation (standard deviation or range chart). A range or standard deviation chart for the so-called "subgroups of two" will show only whether the within-batch variation has changed. The between-batch variation could skyrocket, and the R or s chart would never detect it.

Batch processes are common in the microelectronics industry, and Harris Semiconductor has long experience with them. Examples include tube furnaces and metalization chambers that process silicon wafers. Since gas concentrations and temperatures can vary systematically down the tube's length, we must also be alert for multivariate systems (Levinson 1994b).

Figure 12.19 shows metalization domes and a metalization chamber. Wafers go into the domes, which go into the chamber. The chamber

Figure 12.19. Metallization domes and reactor (batch process).

deposits metal on all the wafers. A single wafer is not an independent representation of the process, because there is load-to-load variation. Figure 12.20 shows boatloads of wafers waiting to go into a tube furnace. We expect variation between furnaceloads, so a wafer is not an independent representation of the process. Figure 12.21 shows wafers waiting to go into a plasma reactor. Again, this is a batch process.

Figure 12.22 shows wafers on a photoresist spin coating track. Pools of photoresist are visible on the wafers. The chucks under the wafers will spin rapidly and cause the solution to spread across the wafers. The wafers then go into ovens that bake off the solvent to leave a film. This equipment coats one wafer at a time, so a single wafer is an independent representation of the spin coating process.

In other industries, batch processes may include furnaces and heat treatment ovens. Mountaintop's standard practice is to look for nested variation sources: between batch and within batch. In the example, two pieces per batch are not a rational subgroup. Two *batches*, however, would be a rational subgroup. Take at least two measurements per batch to estimate $\sigma_{between\ batch}$ and $\sigma_{within\ batch}$. Levinson and Tumbelty (1997) and Montgomery (1984) show how to use one-way ANOVA to isolate the variance components. For three or more stages of nesting, Manugistics' StatGraphics can quantify the variance components.

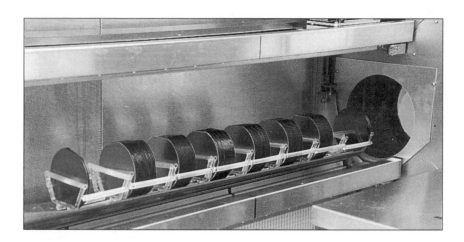

Figure 12.20. Wafers in front of a tube furnace (batch process, multivariate system).

Figure 12.21. Wafers and plasma reactor (batch process).

Figure 12.22. Wafers on spin coating track (non–batch process).

Figure 12.23 (Levinson and Tumbelty 1997, 161) shows Figure 12.16 with the correct control limits. The control limits include within- and between-batch variation.

Nonnormal Systems

Traditional SPC relies on the assumption that the process data follow a normal distribution. In real situations, this is often a poor assumption. Many people get into trouble when they buy "instant SPC" software. They throw their data into a software package that automatically treats the data as normal, but they aren't normal. Mountaintop's statistical training programs stress the need to check the data for normality.

Figure 12.24 shows a histogram of chemical purity data from one of Mountaintop's suppliers. These are real data, although they have been coded to protect the supplier's confidentiality. The bell curve has the data's average and standard deviation. What's wrong with this picture?

First, the bell curve doesn't fit the data very well. Second, part of the bell curve's area is below zero. This suggests the chance of getting less than

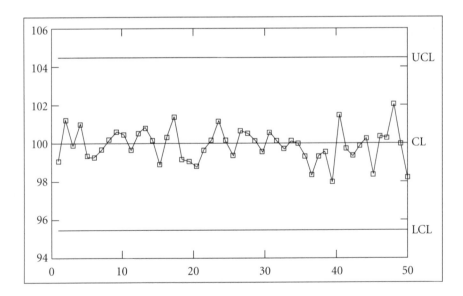

Figure 12.23. Chart for process average, nested variation sources (correct).

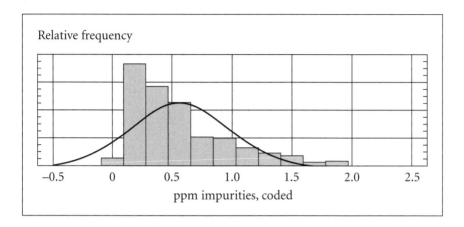

Figure 12.24. Chemical purity data (actual supplier data, coded).

zero impurities, which is impossible. Conclusion: The data don't come from a normal distribution! The assumption that they do will produce misleading results. However, most commercial SPC software makes exactly this assumption.

Chemical manufacturers and chemical users often face this application. Chemical purity is a key quality parameter. There is an upper specification for impurities, but rarely a lower specification; the customer wouldn't object to zero impurities! Microelectronics manufacturers also have one-sided electrical specifications for many products.

A one-sided specification is a hint that the data may not come from a normal distribution. Treating the system as normal may lead to grossly incorrect (by orders of magnitude) estimates of quality. In Figure 12.24, the normal assumption estimated that 73.9 lots per million will exceed a 2 ppm specification. The data actually fit a three parameter gamma distribution, and 7.19 lots per thousand exceed the specification. The outgoing quality (7190 nonconforming lots per million instead of 73.9) is almost 100 times worse than the normal distribution predicts!

In summary, *always check the data for nonnormality.** The chi square goodness-of-fit test, normal probability plot, and Kolmogorov-Smirnov (KOSM) test are common techniques. If the data come from a nonnormal system, Lawless (1982) provides methods for fitting them to other distributions. After fitting the data, be sure to test them for goodness of fit to the selected distribution!

Useful References

A single chapter cannot begin to cover all aspects of industrial statistics. The following references are useful. Some are available from ASQ Quality Press as well as from their original publishers.

ASTM. 1990. *Manual on presentation of data and control chart analysis,* 6th ed. Philadelphia: American Society for Testing and Materials.

AT&T. 1985. *Statistical quality control handbook,* 11th printing. Indianapolis, Ind.: AT&T Technologies. (Formerly known as the *Western Electric Handbook.*)

Barrentine, Larry B. 1991. *Concepts for R&R studies.* Milwaukee: ASQC Quality Press. (How to do R&R studies for gages. Includes the General Motors "long-form" method.)

Holmes, Donald. 1988. *Introduction to SPC.* Schenectady, N.Y.: Stochos Inc., 518-372-5426. (A good, concise reference for SPC practitioners.)

Hradesky, John. 1988. *Productivity and quality improvement: A practical guide to implementing statistical process control.* New York: McGraw-Hill. (A useful, comprehensive reference.)

Hradesky, John. 1995. *Total quality management handbook.* Milwaukee: ASQC Quality Press. (Chapter 8, "Core Techniques," is a useful reference.)

---

*We sometimes say, "Check them for normality," but we can never prove that a population is normal. Tests for normality all begin by assuming normality (like the presumption of innocence in a trial). They then reject this assumption if there is enough evidence of nonnormality. Acquitting a defendant does not prove innocence, and failure to reject the normality assumption does not prove the distribution normal. Nonnormal distributions can pass goodness-of-fit tests, although highly nonnormal systems are less likely to do so.

Juran, Joseph, and Frank Gryna. 1988. *Juran's quality control handbook,* 4th ed. New York: McGraw-Hill. (A comprehensive handbook for quality management systems. It does not get deeply into statistics, but it covers all the basics. A vital reference for anyone who is taking the American Society for Quality's certification exams in quality engineering, quality management, reliability engineering, or quality auditing.)

Lawless, Jerry. 1982. *Statistical models and methods for lifetime data.* New York: John Wiley & Sons. (A comprehensive reference for fitting data to nonnormal distributions. Lawless shows, for example, how to fit data to two and three parameter Weibull and gamma distributions.)

Levinson, W. 1994. Statistical process control in microelectronics manufacturing. *Semiconductor International,* November, 95–102. (The article discusses nested sources of variation in batch processes, and multivariate models. Montgomery (1991) shows how to set up control charts for multivariate systems.)

Levinson, W. 1995. How good is your gage? *Semiconductor International,* October, 165–170. (The article shows how gage capability affects the process capability indices, and the power of SPC charts. The article also shows how gage variation creates the risk of accepting bad pieces or rejecting good ones.)

Levinson, W. 1996. Do you need a new gage? *Semiconductor International,* February, 113–118. (The article describes how gage capability affects outgoing quality.)

Levinson, William, and Frank Tumbelty. 1997. *SPC essentials and productivity improvement: A manufacturing approach.* Milwaukee: ASQC Quality Press. (Most of the book uses diagrams and simple math to teach readers about variation, accuracy, and control charts. The book also covers histograms, cause-and-effect diagrams, flowcharts, ISO 9000, and other quality tools at a high-school level. There is an advanced chapter on setting up control charts, gage studies, and process capability indices; technical appendices that show how to set up multiple attribute charts, perform gage studies, and handle nested sources of variation; and examples of how to use spreadsheets in SPC. There is an optional set of transparency masters for teachers, professors, or company trainers to use.)

Messina, William. 1987. *Statistical quality control for manufacturing managers.* New York: John Wiley & Sons

Moore, David S., and George P. McCabe. 1989. *Introduction to the practice of statistics.* New York: W. H. Freeman and Company. (One of the book's reviewers recommended this as a good introductory—high school, elementary college—textbook.)

Montgomery, Douglas. 1984. *Design and analysis of experiments,* 2d ed. New York: John Wiley & Sons. (A fourth edition is available.)

Montgomery, Douglas, and Elizabeth Peck. 1982. *Introduction to linear regression analysis.* New York: John Wiley & Sons.

Montgomery, Douglas. 1996. *Introduction to statistical quality control,* 3rd ed. New York: John Wiley & Sons.

CHAPTER 13

The Internet

William A. Levinson

"We can get on the nets as full-fledged adults, with whatever net names we want to adopt, if Father gets us onto his
citizen's access."
"And why would he do that? We already have student
access. What do you tell him, I need citizen's access so I can
take over the world?"

—Orson Scott Card, *Ender's Game* (1985)

Here is a one-sentence summary of this chapter's message: *Ride the
Information Superhighway, or be roadkill.* The chapter will start with a general overview of the Internet and some projections on how it will change
the world. Next, it will show how Harris is using the Internet and give
some examples of how you can use it.

The Internet is the third major advance in communications technology; the printing press and the telegraph were the first and second. Three
*I*s describe the Internet and its ability to share information: *instantaneous,
international,* and *interactive.* The Information Superhighway will create
an apocalyptic crisis for businesses and other organizations during the
next 15 years. *Crisis* means "danger" and "opportunity." Those who exploit
the Internet will prosper, and those who don't will be grease spots.

Harris Semiconductor is using the Internet to serve customers and improve quality. Mountaintop is investigating new developments in browsers and supporting software, and how these can help deliver information and services. Hypertext markup language (HTML) is the Internet's standard building block. A Harris Web page provides a detailed example of HTML and shows how it affects the page's appearance and function.

Rapid Technological Change: A Perspective

In Orson Scott Card's (1985) *Ender's Game,* two exceptionally gifted teenagers use a worldwide computer network to manipulate public opinion. The world finally elects the brother as its ruler. Card published the story in 1985, when few people used the Internet. In those days, an Intel 80286 (16-bit) processor was "fast," a 20 megabyte hard drive was "large," and a megabyte was "a lot of RAM." The setting of *Ender's Game* is the distant future,* but the story's computer technology exists today. No one can yet use it to control public opinion, since most people still watch television and read newspapers. However, cable technology may soon allow a television to operate from a computer network. The Internet will change forever the way we exchange information.

The Internet as a Crisis: Danger and Opportunity

And on the pedestal these words appear:
"My name is Ozymandias, king of kings:
Look on my works, ye Mighty, and despair!"

Nothing beside remains. Round the decay
Of that colossal wreck, boundless and bare
The lone and level sands stretch far away.

—Percy Bysshe Shelley, *Ozymandias*

---

*The predictions of science fiction writers often come true far sooner than even the writers expect. Isaac Asimov's (1958) *The Feeling of Power* placed four-function hand calculators hundreds of years in the future. His last *Foundation* books predicted handheld book players (for example, books on CD-ROM) and talking computers.

The Chinese characters for *crisis* mean danger and opportunity. This describes the competitive environment of the late twentieth century, and it includes the Internet. Organizations—even large ones—that ignore change can easily suffer Ozymandias' fate. Even the late nineteenth century has examples, and the late twentieth century offers more.

The Internet's capability for instant, graphical, and interactive communication is turning every marketplace and political arena into a free-fire zone. Within this zone, an organization has three options. It can retreat (get out of the market or business). It can advance through the zone as quickly as possible; both Alexander the Great and General Patton endorsed this approach.* In business, this means embracing and exploiting the changing technology. The third option is to do neither, stand like a deer in the oncoming headlights, and get chopped to pieces.

Organizations need aggressive, dynamic, and adaptive characteristics to succeed in this environment. Bill Gates (1995) says that the Internet will lower or remove entry barriers and create a frictionless marketplace. *Frictionless* means that distribution costs will consume much less of the product's or service's value.

This competitive environment is an open, flat, accessible playing field—or battlefield. There is no barrier to entry or movement, no cover, and nowhere to hide. The implications are clear for organizations that rely on "Maginot Lines" of entry barriers or distribution channels to protect their market shares.

Organizations Must Anticipate Change and Adapt to It

Cohen and Gooch (1991, 24–28) describe three errors that organizations make.

1. Failure to learn from past experience (one's own, or others')

2. Failure to anticipate (lack of foresight, proactivity)

3. Failure to adapt to a new situation

---

*Patton said that the effectiveness of the enemy's fire depends on (1) its intensity and (2) the length of one's exposure to it. Cover reduces its intensity, but rapid movement limits one's exposure. Alexander the Great defeated the Persians by charging through their archers' missile zone as rapidly as possible, thus limiting his troops' exposure.

Simple failure results from a single error or deficiency; adaptable organizations often recover from simple failure. Complex failure or aggregate failure is the combination of two kinds of failure. Catastrophic failure results from the simultaneous or consecutive operation of all three mistakes: failure to learn, failure to anticipate, and failure to adapt. It often results in the organization's destruction.

> *This "telephone" has too many shortcomings to be seriously considered as a means of communication. The device is inherently of no value to us.*
>
> —Western Union internal memo, 1876

Western Union was in the electronic communications business that, in 1876, meant telegraphy.* This memo suggests that Western Union's self-limiting paradigm was that telegraphy was the only possible form of electronic communication. The company had a chance to buy Alexander Graham Bell's telephone patent, but turned it down. On March 3, 1885, the American Telephone and Telegraph Company incorporated as a subsidiary of American Bell Telephone Company. In 1911, AT&T absorbed Western Union in a hostile takeover.†

Microsoft CEO Bill Gates (1995) provides ample warning of what happens to organizations that don't pay attention to technological change. Microprocessors were interesting toys for hobbyists before the appearance of operating systems. Gates recognized the value of combining microprocessor technology with an operating system, and he dropped out of Harvard to seize the opportunity. The introduction of operating systems turned microprocessors into commercially viable products that are transforming every aspect of business and modern society.

---

*The telegraph, which uses dots and dashes (ones and zeros), was the conceptual forerunner of the modem. A skilled operator could send or receive information at a few baud, or bits per second. A modern 28.8 K modem is about 10,000 times as fast.

†"A History of the Telephone," prepared by Dawne M. Flammger, February 1, 1995 at http://www.geog.buffalo.edu/Geo666/flammger/tele2.html, describes the takeover of Western Union by AT&T. (Note: This address was valid in April 1997, but addresses are subject to change or removal. Some Internet references may be unavailable when this book goes to press or may become unavailable later.)

Personal computers did to the mainframe computer what automobiles did to passenger trains. IBM itself helped popularize the technology that was to be its undoing, by introducing the IBM Personal Computer. "IBM compatible" standards, and Microsoft's operating systems, helped spread this technology. IBM did not make a mistake by creating the personal computer since, if it hadn't, someone else would have. Apple had already been making minicomputers for a few years. IBM, however, did not learn from others' past experience—specifically, the railroads'.

A modern automobile is usually more convenient than a train. When early automobiles appeared, this might not have been true. Cars were not as fast or reliable as they are now, and roads were not as good. Nonetheless, the railroads needed to recognize that they were in the transportation business, not the railroad business. The Interstate highway system, and the passenger airplane, put most passenger trains out of business. However, commuter trains still enjoy a niche market near big cities, where traffic conditions make driving unpleasant and inconvenient.

A minicomputer is far more convenient than a mainframe, and computer technology advances at a phenomenal rate. IBM was in the information processing business, not the mainframe business. Despite its legendary marketing team and its legions of MBAs, IBM did not recognize the personal computer's implications. This is why the computer market left IBM behind, while Bill Gates is the richest person in the United States.

This was not the first time that paradigms almost caused IBM to miss a big market shift. In 1943, Thomas Watson Sr. said, "I think there's a world market for about five computers." Watson, however, could apparently change his mind or listen to good advice. Technological change was not as rapid in the 1940s, and the competitive environment was more forgiving. In the late 1980s and early 1990s, however, Watson Sr. was long dead and his son had retired. IBM missed the opportunity—and the turbulent, dynamic competitive environment of the late twentieth century does not forgive such errors.

Three other big companies—Digital Equipment, Atari, and Hewlett-Packard—also missed the personal computing opportunity. In 1973, Digital's Ken Olson said, "There is no reason for any individual to have a computer in their home." Atari, and then Hewlett-Packard, refused to hire Steve Jobs and Steve Wozniak. HP said, "We don't need you. You haven't [finished] college yet." Jobs and Wozniak decided to do it themselves, and they founded Apple Computer. We are not talking about stodgy, tradition-bound, inflexible organizations, either. Innovation is a deeply ingrained part of Hewlett-Packard's culture, while IBM dominated the computer market because of its daring gamble on the System 360. Even innovative organizations, however, can miss a step. The implications of the Internet are so wide and far-reaching that anyone who misses this one is going to be roadkill.

The Internet: The Third Major Advance in Communications

As already stated, the three major advances in communications are the Gutenberg press, the telegraph, and the Internet. Here is a summary of their effects on society, technology, and business.

1. Fifteenth century: The Gutenberg press allowed mass production of written material.

   —Education was now within the reach of ordinary people, since they did not have to buy hand-copied books at a frightful cost. This also broke the Catholic Church's monopoly on education (the monks copied the books) and reduced its political influence.

   —Without the Gutenberg press, Martin Luther could not have launched the Reformation.

   —Mass production of books helped science advance far more rapidly.

2. Nineteenth century: The telegraph allowed instant point-to-point communications.

   —This development coincided with the growth of the railroads and permitted the coordination of train movements. The railroad, in turn, helped the United States expand across North America.

3. 1990s: The Internet delivers instant, international, and interactive communications at the convenience of those who wish to receive them.

The Gutenberg Press: Mass Production of Books. In the fifteenth century, the Gutenberg press allowed the mass production of printed material. Before the invention of moveable type, scribes had to copy books by hand. Books were rare, costly, and inaccessible to the masses. The Catholic Church, whose monks copied the books, had a virtual monopoly on knowledge and education. Illiteracy was rampant, even among the upper classes. In Mark Twain's (1989) *A Connecticut Yankee in King Arthur's Court*, an Arthurian noble considers the suggestion that he can read to be an insult. Such work, he says, is for clerks (or clergymen) and the like.

The Gutenberg press changed this by making books and other printed material accessible to everyone. It allowed information to spread through Europe, albeit only as quickly as a horse could carry it. Nonetheless, it caused radical changes in society and it promoted technological change.

The Telegraph: Instant Point-to-Point Communications. Electronic communications, in the form of the telegraph and telephone, were the second major advance. Even in the early nineteenth century, information could travel only as quickly as a horse or ship could carry it. Andrew Jackson fought the British at New Orleans after the War of 1812 was over, because neither he nor the British commander had received the news. The telegraph changed this and promoted the growth of the United States by linking its Pacific and Atlantic coasts. The telephone improved on the telegraph by allowing voice communications, and communication satellites then removed the need for undersea cables.

The telephone, and then the fax, merely improved on the telegraph. They are better forms of electronic, point-to-point communications than the telegraph, but they are still point-to-point.

The Internet: Instantaneous and Interactive Mass Communication. The Internet is the third major change. It is not a mere evolution or improvement of other media, but a revolution. To survive and prosper, organizations must recognize this, and use this new technology.

The Internet allows low-cost, instantaneous, and widespread worldwide communications. If you wanted to talk with someone in Australia or India, you had to (1) find him or her, (2) place a costly international call, and (3) make sure your working-hours call didn't reach your party at two or three in their morning. Now one can use an online directory to find a person or organization and send electronic mail. Online newsgroups are accessible by anyone, anywhere in the world.

The Internet allows its users to search for and retrieve massive amounts of information. This chapter will later discuss online literature searches and search engines.

The Internet is interactive. Web pages on interactive servers can receive and process information. They can, for example, act as automated sales clerks.

Implications for National and International Politics

The Internet will play a major role in national politics. It also crosses international boundaries and is almost uncontrollable. It will play a major role in displacing totalitarian governments. The role of the printing press in the Reformation (sixteenth century C.E.) predicts this,* and the next section discusses it.

The Internet and Freedom

It's unlikely that anyone will use the Internet to take over the world. Instead, the Internet will free many countries from oppressive governments.

1. The Internet crosses international boundaries and is almost uncontrollable.

2. Dictators and tyrants must control what their people read and hear.

   —To prevent their people from reading politically incorrect material, they must ban computers (or at least modems).

   —To use the Internet for technological advancement, they must encourage their people to use computers and modems.

If dictators forbid their people to use the Internet, their countries will fall behind economically and technologically. They will be unable to develop military forces that can threaten an advanced democracy or republic. If they let their people use the Internet, they will lose control over them. The

---

*See http://www.strangelove.com/publish/paradigm/chp1.html, "Understanding Internet: The Democratization of Mass Media and the Emerging Paradigm of Cyberspace" by Michael Strangelove.

totalitarians will have to choose between control and progress; they can't have both.

A world map (Borman 1996) supports this contention. Iraq, Syria, Libya, Burma, and North Korea lack Internet or e-mail connectivity. These are among the most abusive and repressive dictatorships in the world. Among dictatorships, only Iran and Cuba have Internet connectivity. In contrast, Europe (except the former Yugoslavia), North America, and most of Latin America have Internet connectivity. Russia, Egypt, Israel, Turkey, Pakistan, India, Indonesia, Japan, Australia, New Zealand, Taiwan, and Malaysia also have access to the Internet. In summary, nations with any semblance of "rule by the consent of the governed" have Internet access.

The conflict between Martin Luther and the Catholic Church was a harbinger of this situation. Luther could not have spread his teachings without mass communications, and the Gutenberg press made this possible. The press allowed the printing of non-Latin prayerbooks, which allowed the masses to worship without a priest. The Church's reaction shows the effectiveness of mass communications. In those days, people literally flamed their religious opponents, but the Church could not get its hands on Luther to burn him. (To "flame" someone on the Internet means to insult or criticize, often in a public forum like a newsgroup.) The Church had to content itself with burning Luther's books and using the printing press to publish bulls (bulletins) of excommunication against him. Luther, in turn, reacted by burning these. Both sides used mass communications to advance their positions, and each side feared such communications by the other.

Now recall that the printing press was the first major advance in communications. Electronic communications were the second, and the Internet is the third. Unless totalitarian governments close their borders to electronic communications (with the attendant negative effects), the Internet will overthrow them.

Politicians Are Still Getting on Board

There was a saying, "Never pick a fight with someone who buys ink by the barrel." Today, it is "Never pick a fight with computer-literate people who can communicate with each other by pushing a button." The U.S. Congress passed the Communications Decency Act (CDA), but in 1997 the Supreme Court ruled it unconstitutional. The Internet allowed the bill's opponents to unite quickly and mobilize opposition to the legislation.

Independent public interest groups such as Vote Smart (http://www.vote-smart.org/) are using the Internet to change the political process.

> *Project Vote Smart's goal is simple: turn the same slick, high technology used by candidates and elected officials around to the citizen's advantage instead. Allow each individual citizen to decide for himself or herself what information is important and then provide instant access to it. In our democracy no power is more decisive, no political weapon more potent, than dependable, accurate, factual information.**

Implications in Business

Memo

To: The Frogs Sitting in the Pot on the Stove

Subject: Your Prospects for Long-Term Survival

Dear Frogs,

The water (or market segment) in which you are sitting is only lukewarm today. There is, however, a big fire under it, and it's called the Internet. If you don't start hopping soon, you will be dinner.

---

*"Permission to Use Project Vote Smart Data: Project Vote Smart does NOT copyright any of its publications or databases. We do this so that citizens, schools, libraries, and other users may copy and distribute the information freely."

It's not likely that two teenagers, no matter how intelligent, will use the Internet to rule the world. It's very likely that people will use the Internet to rule the marketplace, or at least the market segments they want to rule. The World Wide Web will bring unprecedented change to the business world. Table 13.1 shows some examples.

Jeffrey Christian, who runs an executive search firm in Cleveland, Ohio, says, "The Internet is a rifle aimed at all middlemen: insurance salespeople, investment bankers, travel agents, car dealers. It's going to hit everybody" (Zachary 1996). The same article discusses banks that are reducing personnel with automated systems. Security First Network Bank of Pineville, Kentucky, is doing business *only* on the Internet. Security First's CEO, James S. Mahan, says "It means we don't need all these bodies." Gates (1995) delivers a similar message.

Intranets

> *Our competitors won't know what hit them.*
>
> —A manager at Weyerhaeuser Corporation

Intranets, or internal computer networks, make businesses *porous* to customers and suppliers. They support just-in-time manufacturing and reduce staff requirements, thus reducing product costs.

W. Edwards Deming said to break down organizational barriers inside companies. Tom Peters (1988) extended Deming's advice by saying to make the organization porous to customers and suppliers. It is easy for customers and suppliers to do business with a porous organization. In contrast, they must leap bureaucratic hurdles to work with a nonporous one. Intranets, or internal computer networks, remove internal and external obstacles.

Richards (1996) writes,

> *These internal networks usually start out as ways to link employees to company information, such as lists of product prices or fringe benefits. Then, the networks are gradually extended to major suppliers and customers. The next step, which few companies have taken but which will bring them the biggest change, is to allow these three groups—employees,*

Table 13.1. Industries and services subject to displacement by the Internet.

Frog (Industry/ Service)	Impact of the World Wide Web
Postal service	Electronic mail had the limitation that it could carry only text. New e-mail systems, such as Netscape Navigator 3.0 and Microsoft Internet Explorer, allow attachment of files, including color images and sound files. The Internal Revenue Service now allows the electronic filing of tax returns. Compare 32 cents for delivery in a few days, versus instant delivery for free. An Internet account requires a monthly service fee, but there is no marginal cost for sending e-mail. Spammers, or people who abuse the Internet with mass mailings of bulk e-mail, take advantage of this aspect. The Internet will never displace the postal service, since some items (greeting cards, written contracts, packages) cannot be e-mailed. However, electronic mail will erode the first-class mail segment.
Long-distance telephone carriers	Software and hardware are available that allow people to talk over the Internet. Both parties must have the equipment and software. If they do, however, one can call New York from Tokyo for the price of a local call.
Executive recruiters, help-wanted classifieds	Companies are posting jobs on electronic bulletin boards and Web sites. As more people become Internet-literate, companies will not have to pay recruiters or buy advertising space. Recruiters may own and manage Web sites, but there will be fewer jobs in this profession.
Stockbrokers	There are already plenty of online discount brokerages. E*Trade Securities (http://www.etrade.com) charges a flat $15 to $20 commission for trades of up to 5000 shares. (See also Charles Schwab online at http://www.schwab.com, Lombard Institutional Brokerage Inc. at http://www.lombard.com, and PAWWS Financial Network, http://www.pawws.com.) Andrew Klein of Spring Street Brewing says, "Brokerage fees on the Web are coming down, and will eventually near zero because the cost of each additional transaction is so minuscule" (O'Connell 1996).
Car dealers (sales organization)	Car salespeople had better start looking for real jobs. There are already online car brokers that cut the salesperson out of the loop. One buyer saved $4000 off the list price by using Auto-By-Tel (http://www.autobytel.com) (Wiener 1996). Levinson (1994c, 204–206) recommends just-in-time factory ordering, without a car dealer as intermediary. The author stands by this prediction.

> *suppliers, and customers—to conduct business on the*
> *Internet in a completely automated way. Customers would*
> *buy products, suppliers would be paid and employees would*
> *gain access to the company's database, all through a single*
> *computer network.*

You can't get much more porous than this.

Chapter 9 of this book discusses synchronous flow manufacturing (SFM). SFM is similar to just-in-time manufacturing; it ties production starts to the process constraint. Instead of pushing product into the line, SFM pulls it in, with the constraint setting the pace. Harris Semiconductor has used SFM to slash its inventories and increase throughput. At Mountaintop, the slowest manufacturing operation is the constraint, since the factory can sell everything it makes.

What if the marketplace is the constraint? We cannot sell everything we make, and we would prefer not to make inventory. Weyerhaeuser Corporation, a major pulp and paper manufacturer, is using its Intranet to manage its shipments to a customer's factory. The system takes the customer's sales information and links it to Weyerhaeuser's Valley Forge Fine Paper Company's master production schedule. This is intercompany SFM, with the customer's factory setting the pace. As the customer uses the Weyerhaeuser products, the Valley Forge mill schedules more. The system reduces the customer's inventory by up to 40 percent. It also allowed the Valley Forge mill to get rid of some warehouse space (Richards 1996). Since inventory ties up cash, the system improves both companies' cash flow and liquidity.

A senior Weyerhaeuser manager says, "We're just-in-timing every aspect of the manufacturing, ordering, and delivery process." The company hopes the system will "double the unit's productivity without adding any additional staff, allowing it to undercut competitor's prices and grab additional market share" (Richards 1996).

The Intranet at Harris. Mountaintop is examining ways to use an intranet for process documentation. A single HTML document can display a process flowchart and include links to each operation's documentation. This provides a convenient overview of an entire process and allows easy access to its process capability studies, failure mode effect analyses, gage studies, and so on.

Harris Semiconductor plans to let the self-directed work teams post their meeting times, minutes, and projects on the intranet. This supports the teams by making it easy for everyone to know where and when to attend the meetings.

Retailers Pay Attention (Anyone Want a Used Shopping Mall?)

Don't worry, it's only 0.021 percent of the market.

—heard from those about to become grease spots
on the Information Superhighway

A consultant performed a job for a wealthy company (or kingdom), which asked him to name a price. The consultant said, "Take a chessboard with 64 squares. Place one penny on the first square, two on the second, four on the third, and so on until the board is full. That is my price." It was before the days of calculators and computers, so the client agreed.

The client later discovered that the cost was 1.84×10^{17} (2^{64} cents minus one cent), or more wealth than the world has created in its entire history. This is an example of a 100 percent growth rate, and the Internet retail market comes close to this.

In 1995, sales over the Internet totaled $350 million, or 0.021 percent of the retail industry's $1.7 trillion. Here are some projections for online shopping revenue (Sandberg 1996).

Year	Dollars (billions)
1995 (actual)	0.350
1996	0.518
1997	1.138
1998	2.371
1999	3.990
2000	6.579

Figure 13.1 shows a semilog plot of these numbers. The slope is 0.267, which suggests an annual growth rate of $10^{0.267} = 1.85$, or 85 percent. The six points have a 0.996 correlation, and the figure for 2010 is $3.3 trillion. While extrapolation is always dangerous,* the handwriting is on the wall. The national economy is not going to grow 85 percent a year, so the only way for Internet sales to continue this growth rate is to take sales from traditional retailers.

Mene, Mene, Tekel, Upharsin: You have been weighed in the balance, and are found wanting. Your market share is divided, and given to the Medes, Persians, and anyone else who has an Internet presence.

Frog legs? We're sitting in this cozy pot (although it is starting to get a wee bit warm), and someone is talking about frog legs. It'll never happen—we own this market segment, don't we?

---

*Executives at the Internet Commerce Exposition in Atlanta projected $2 billion in Internet commerce for 1997. "Optimists believe Web-based dealings could reach the $1 trillion mark by the year 2010," while Wells Fargo and Company says it picks up 20,000 new online customers every two weeks ("Web begins to demonstrate value to business," Reuters News Service, April 1997).

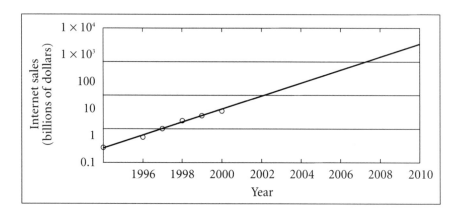

Figure 13.1. Internet sales.

International Conferencing: From Your Home or Office

Prediction: Traditional conference centers and hotels will be obsolete by the year 2010. The American Society for Quality's 65th Annual Quality Conference will take place on the Internet.

Don't invest in a conference center or hotel. Except for those that offer exotic vacations to go with business meetings (for businesses that can afford them), they will be obsolete. A press release from IBM Corporation* described the virtual reality markup language (VRML) and its applications. These include the following:

1. Virtual branch offices and conference rooms, in which avatars, or three-dimensional representations of the conferees, meet

2. The IBM Digital Library

3. IBM's new product gallery

International Sightseeing Tours for $100 (or Less)

Take a guided tour of the Vatican, Sistine Chapel, Kremlin, and the Louvre for less than $100. Talk to visitors from other countries while you're there. Don't worry about crowded airplanes, customs, passports, jet lag, or bad hotels. The tour starts and ends in your own home.

This is an obvious extension of virtual reality and the Internet. A company could begin by taking extensive professional photographs of historical sites and converting them to digitized images. Professional tour guides could add written and audible commentary. The vendor could supply this on CD-ROM—and such products probably exist today. The next step is to place it on the Internet, so visitors from all over the world could take the "guided tour" together. Their avatars could, for example, tour the Louvre and Notre Dame Cathedral, then meet in a Paris cafe for lunch.

---

*http://www.software.ibm.com/press/vrml.html, but the link is no longer available.

This service would have educational applications, too. Students who are learning a foreign language could "visit" the country, hear native speakers, and learn the culture and customs. This could also be a valuable experience for businesspeople who plan to visit (virtually or in person) another country.

The End of the City?

The Internet, and telecommuting, could easily make cities obsolete. Cities evolved thousands of years ago to meet two specific needs: security and centralized commerce. A city was easy to defend if its inhabitants built a wall around it, and most cities had central marketplaces.

Today, cities have traffic congestion, high living costs, high taxes, and often high crime rates. Environmentalists are complaining about air pollution, and they want cities to limit automobile traffic. Cities no longer provide any military benefit; the Second World War showed them to be convenient targets for air raids. Land in cities is so expensive that no one is likely to build a new factory in one. Urban businesses include service industries like banking and finance, entertainment, and retail shopping.

Knowledge workers, such as the people who work in the banking and finance industries, can often perform their jobs wherever a computer is available. They do not need to live in the city or even drive into it. Meanwhile, city stores have to pay exorbitant rents, high taxes, and other overhead costs. Suburban malls have already crushed many downtown retail districts, and online shopping will only accelerate this trend. Nothing, however, can replace a live Broadway performance or a live opera. These may be the only urban businesses that will survive the oncoming tide of change.

We've seen a preview of where the Information Superhighway will take us in the twenty-first century. The message is clear: Ride it or be roadkill. Now we'll look at where it is today and how it can help us.

Internet Resources

The Internet is an indispensable resource for technical research. In the 1970s, a chemical literature search was an arduous process. I remember spending hours in the chemical libraries at Penn State, and later Cornell, looking through volumes of *Chemical Abstracts*. After finding potentially useful references, one had to look for them and hope they were in the library. There was a good chance of finding them in large university libraries, but most business organizations do not have huge libraries.

Today, one can search research journals by pushing a button. CARL Corporation (http://database@carl.org/cinfo.html) offers free online literature searches and delivers articles by fax for a fee. The fee includes CARL's service charge and the copyright holder's fee. (Go to the UnCover home page at http://uncweb.carl.org/.)

One can also use Internet search engines to look for Web pages. Digital's Alta Vista (http://www.altavista.digital.com/) is among the most powerful. Its advanced search option allows the user to place logical conditions on the search, such as "and," "not," and "or." Alta Vista and a couple of other search engines allow the user to place words in quotation marks. A search for "discrete power" requires an exact match to these words. The presence of capital letters also requires an exact match. Lowercase letters (in Alta Vista) allow either case. Less sophisticated search engines will return documents that contain "discrete" and "power" anywhere.

A search for +EWMA +SPC yielded 55 links to pages with information on exponential weighted moving averages and statistical process control. The links include companies that sell SPC software, university courses, and pages with free information.

Standards and Regulations
Standards and government regulations (for example, those of OSHA) are available through the Internet (see Table 13.2).

Professional Organizations
Most professional organizations have Web pages on the Internet. Table 13.3 shows some professional organizations with Internet web pages.

Table 13.2. Standards and regulations on the Internet.

Resource	Link (as of July 1997)
American National Standards Institute	http://www.ansi.org/
American Society for Testing and Materials	http://www.astm.org/stand.html
FedWorld Information Network	http://www.fedworld.gov/
MSDSs online (Material Safety Data Sheets)	http://129.252.111.69/hs/msds.htm
National Institute of Standards quality programs (Baldrige Award)	http://www.nist.gov/quality_program/
NIST Standard Reference Data Products Catalog	http://www.nist.gov/srd/srd.htm
NIST Technology Services Headquarters	http://ts.nist.gov/ts/htdocs/200.html
Occupational Safety and Health Agency	http://www.osha.gov/
QS-9000	http://www.asq.org/9000/
U.S. Patent and Trademark Office	http://www.uspto.gov/

Table 13.3. Professional organizations on the Internet.

Organization	Web Site (as of July 1997)
American Chemical Society	http://www.acs.org
American Institute of Chemical Engineers	http://www.aiche.org/
American Society for Quality	http://www.asq.org/
American Society of Civil Engineers	http://www.asce.org
American Society of Mechanical Engineers	http://www.asme.org
ASM International (materials and metals science)	http://www.asm-intl.org/
American Statistical Association	http://www.amstat.org/
Institute of Electronic and Electrical Engineers	http://www.ieee.org
National Academy of Sciences	http://www.nas.edu/
National Society of Professional Engineers	http://www.nspe.org/
Society of Manufacturing Engineers	http://www.sme.org/
TechExpo directory of technical, engineering and science organizations	http://www.techexpo.com/ tech_soc.html

Company Visibility

The Internet is becoming an online buyer's guide. Customers can search the Web for products that meet specific needs. Unlike a buyer's guide, a Web site can provide extensive, detailed information about a company's products.

This is your company in a traditional buyer's guide (or trade magazine). Remember, you're paying by the inch, page, or other unit of space. It's black and white, unless you pay extra. Your prospective customer has to buy the guide.

XYZ Corporation
123 XYZ Drive
Widgettown, NY 12345
800-555-1234

We make pumps and valves for chemical applications. We offer centrifugal pumps, diaphragm pumps, and positive displacement pumps. Our globe valves, gate valves, and needle valves come in standard materials or in specialty materials for handling corrosive materials.

This is your company on the Web. If you're paying, you're paying by the megabyte, and you can fit a lot into a megabyte. This includes color photos, animations, video clips, and so on. All your customer has to do is log on.

XYZ Corporation
123 XYZ Drive
Widgettown, NY 12345
800-555-1234

Pumps for chemical applications:
Centrifugal pumps*
Diaphragm pumps
Positive displacement pumps

Valves:
Globe valves
Gate valves
Needle valves

E-mail **TechSales@XYZ.com**

Page on centrifugal pumps
• Model numbers and color pictures
• Specifications
• Prices
• Performance data

Page on globe valves
• Model numbers and color pictures
• Prices
• Friction factor data
• List of specialty materials (with links to information on corrosion resistance for each)

Any questions?

---

*Underlined, boldface text refers to a link. Pointing the mouse to **Centrifugal pumps** and clicking takes the user to the page on centrifugal pumps.

Thomas Register (http://www.thomasregister.com:8000/) now has an online, searchable catalog. Some of the entries have links to online catalogs.

Advertising Effectiveness

There is a key difference between Web advertising* and advertising through bulk mail, television, or magazines. To be effective, the latter must reach a customer when he or she wants the product. The customer may have no interest at all, may already have the item, or may have no money for it. If the customer later develops an interest or need or gets the money (for example, in next year's budget), he or she has probably discarded or forgotten the ad.

A customer who visits our Web page has, in contrast, come looking for us. The customer had enough interest to locate us through a search engine or another Web site. A customer who has no interest today might come tomorrow, next month, or next year. We don't have to spend money to advertise to an unreceptive audience, and we don't have to worry about the ad's timing. The Web page is always there for whoever wants it, whenever he or she wants it.

Bulk mail

- About 10 cents per contact.
- Often thrown away unread.
- Even with targeted mailing lists, delivery must coincide with the customer's need and budget.

Television

- Extremely expensive.
- Often ignored, or "zapped" (watcher uses the remote control to change the channel).
- Must coincide with the customer's need and budget.

---

*This refers to making information available on a Web page, as opposed to the unacceptable practice of spamming. Spamming means indiscriminate posting of advertisements to mailing lists or newsgroups. Most Web users find spamming highly offensive, and it violates the acceptable use policies of most Internet service providers.

Internet Web page

- Fixed cost to place on the Internet (as low as $60 a year).

- All contacts ("hits," or visits to the Web page) are from people with at least some interest in our product.

- Available whenever the customer has an interest and has money to buy the product.

- A log of visits can provide valuable marketing intelligence. Which pages did customers visit most often?

Harris Semiconductor's Web Sites

Harris Semiconductor's Mountaintop plant has a Web site at http://www.mtp.semi.harris.com/, with a link to Harris Semiconductor's site at http://www.semi.harris.com/. Table 13.4 shows some resources that are available through each page.

Creating a Web Site

An organization can set up a rudimentary Web presence in 20 hours or less. This section provides some examples. It is not a substitute for a comprehensive reference on writing HTML (or Java), but it covers some key points.

It is easy to create a Web site. The standard language is HTML, which sounds intimidating but is easy to use. Any word processor that can generate ASCII (text-only) documents can write HTML. I created a rudimentary personal Web site, with graphics, in 20 hours and with less than $100 in software: $30 for an HTML guide with an HTML editor on CD-ROM (Savola 1995), and $60 for CorelDRAW 3.0, which can convert drawings into .GIF (graphics interchange format) files. I later purchased Alchemy Mindworks' .GIF Construction Set, which converts a set of .GIF files into an animated series. This downloadable shareware package costs $25. A typical online account costs $20 a month, so for $125 in software and $240 a year, anybody can have an online presence.

Software is available that will create HTML code automatically. Netscape Navigator Gold 3.0, for example, has menu-driven systems for

Table 13.4. Harris Semiconductor resources on the Internet.

http://www.mtp.semi.harris.com/	http://www.semi.harris.com/

1. Product reliability data
  • With links to product data sheets.
  • Saves countless hours of phone calls and faxes, and allows Harris' Asian sales office to fulfill customer requests instantly.
  • Allows dissemination of up-to-date information, instead of circulation of expensive glossy publications that appear only quarterly.
2. Information about Mountaintop, Pennsylvania, with a map showing the plant's location.
3. Construction status and photos of a new eight-inch wafer fabrication plant
  • Customers who visited the site in April said they felt they knew the place before they arrived.
4. Harris e-mail and phone directory
5. Links to other Harris Corporation sectors
  • Government
  • Aerospace/military
  • Health care
6. Links to professional societies
	1. Job opportunities
	2. Product information
	  • Power MOSFETs (Metal Oxide Field Effect Transistors), including links to product data sheets
	3. Archive of new product press releases
	4. Design support
	  • Technical notes
	  • Application briefs
	  • Design software
	  • Customers' ideas for new Harris products

doing this. Corel WordPerfect can publish documents to the Web. Corel Barista creates "documents based entirely on the Java language, without any programming requirements."

If you have your own Web server, this is where to put your Web site. If not, your Internet service provider can provide instructions for creating a

Web site. There should be no extra charge for this unless you need huge amounts of memory. Netcom currently includes 1 megabyte of memory with its $19.95 monthly fee, and Delphi offers 10 megabytes. There are providers who sell Web space for as little as $60 a year, although this does not include a dial-up account—that is, you can create a Web site, but cannot browse the Internet. You will need a standard account to view your own Web site if one of these providers hosts it.

You will probably have to pay more if you want common gateway interface (CGI) capability. CGI allows the Web site to interact with users by, for example, taking orders electronically.

File transfer protocol (FTP) is a common method of posting files to a Web site, although service providers may offer other methods.

Figure 13.2 shows the Web page for Harris Semiconductor's new eight-inch wafer plant in Mountaintop.

Table 13.5 is the HTML document (http://www.mtp.semi.harris.com/raptor.html) that creates this Web page. (The contents have changed since preparation of the manuscript.)

Viewing HTML Source Codes. A good way to learn HTML is to look at other people's Web pages. Select VIEW SOURCE on Netscape Navigator's VIEW menu to see the page's HTML code. Select SAVE AS from Navigator's FILE menu to capture the entire page's code as a file. One can then see the commands that produce the desired results and edit them to meet one's own needs.

Background Color. Select a background color that is easy on readers' eyes. Don't camouflage the page's text against the background, or against the wallpaper!

The background attribute command is <BODY BGCOLOR=#RRGGBB>, where RR, GG, BB are hexadecimal color portions for red, green, and blue. CorelDRAW 3.0 can show various colors, and it reports their compositions in percent red, green, and blue. After identification of a good background color, multiply the percentages by 255 (FF) to get the correct hexadecimal numbers. BGCOLOR=#FFFFFF creates a solid white background, while BGCOLOR=#000000 is solid black.

One can also define a text color with <BODY TEXT=#RRGGBB>. <BODY BGCOLOR=#RRGGBB TEXT=#RRGGBB> defines both the

Presents:

World's First 8 inch

Power MOS Semiconductor Fab

In Mountaintop,

MELBOURNE, FL/MOUNTAINTOP, PA, September 27, 1995 -- Harris Corporation today announced it will construct the world's first 8-inch power Metal Oxide Semiconductor (MOS) fab as part of the company's plan to invest $250 million in its power semiconductor business.

The new facility will be a stand-alone, three-story building situated just north of the company's current Mountaintop facility and will include a 30,000- square-foot clean room -- the first of its kind designed specifically to produce power MOS semiconductors on 8-inch wafers rather than the traditional 6-inch wafers. Utilizing the larger wafer dramatically increases the yield - - or number of circuits produced.

Figure 13.2. A Harris Mountaintop Web page.

KEY FACTS

- **PRODUCTS** : MOSFet, IGBT, MCT, and Smart Discretes
- **CLASS** : 100% Auto SMIF - Class 1
- **EQUIPMENT** : State of the Art, IC Class, Smart Tag
- **CAPACITY** : Phase I = 81,000 Wafers/year
- **CAPACITY** : Phase II = 216,000 Wafers/year
- **COMPLETION** : Phase I = January, 1997

- **PROJECT LEADERS : Bill Burrell and Dave Hollock** Construction - UPDATED 7/19/96

Detailed

For more information or Mountaintop job opportunities:
Human Resources
HARRIS Semiconductor
125 Crestwood Rd.
Mountaintop, PA 18707
Phone: 717-474-6761
Other Harris

Clint ◎ - cchamber@harris.com
- For more information about Harris Semiconductors:
  U.S. and Canada callers dial 1-800-4-HARRIS (1-800-442-7747) ext. 700
  International callers dial 407-727-9207

Last modified: 7/19/1996

Figure 13.2. *Continued.*

Table 13.5. Harris Web page: HTML and explanation.

HTML	Explanation
<!DOCTYPE HTML PUBLIC"-//IETF//DTD HTML 2.0//EN"><html>	<! denotes a comment. <html> identifies the document as html.
<head>	Start heading
<META name="keywords" content="wafer power MOSFet IGBT MCT fab 8-inch cleanroom construction photos Pennsylvania Mountaintop Wilkes-Barre" >	Search engines identify this section as offering keywords for indexing.
<title>Raptor Information - HARRIS Mountaintop</title>	Document title
<LINK rev=made href="mailto:cac@mtmis1.mis.semi.harris.com">	
</head><BODY BACKGROUND="raptorbk.gif"> raptorbk.gif	Uses raptorbk.gif as the wall-paper, or background, for the page. (This does not appear in the figure.) One can also specify BGCOLOR (back-ground color) here.
<A HREF="http://www.semi.harris.com/"><IMG SRC="semi_plate.gif" WIDTH=184 HEIGHT=58 BORDER=0 alt="[HARRIS Semiconductor Logo]"></A> semi_plate.gif	This displays the Harris logo semi_plate.gif, and specifies 184 by 58 pixels. If the browser cannot display it, "[Harris Semiconductor Logo]" appears. Clicking on this logo takes the user to http://www.semi.harris.com/.
<p>	Skip a paragraph.
<center>	Center the subsequent material.
<h2>Presents: <IMG SRC="raptor.gif" WIDTH=264 HEIGHT=200 align="middle" alt="Project RAPTOR"></h2> <h1><a name="top">World's First</a> 8 inch</h1> <h1><a href="penn_fab.html">Power MOS Semiconductor Fab</a></h1> <h1> In <a href="mtop.html">Mountaintop, PA</a></h1> raptor.gif	Use heading 2. Headings range from 1 (largest) to 6 (smallest). Display a 264 by 200 pixel raptor.gif, aligned in the middle. If the browser cannot display it, display "Project RAPTOR" instead. Use heading 1 (largest) for the subsequent text. Clicking on **Power** takes the user to penn_fab.html. Clicking on **Mountaintop, PA** takes the user to mtop.html.

Table 13.5. *Continued.*

HTML	Explanation
</center>	End the centered format (</ always means to end or terminate a command).
<strong>MELBOURNE, FL/MOUNTAINTOP, PA, September 27, 1995 — Harris Corporation today <a href="penn_fab.html">announced</a> it will construct the world's first 8-inch power Metal Oxide Semiconductor (MOS) fab as part of the company's plan to invest $250 million in its power semiconductor business.	<strong> emphasizes the text. Click on **announced** to go to penn_fab.html.
<p>The new facility will be a stand-alone, three-story building situated just north of the company's current Mountaintop <a href="HarrisSemi_blurb.html#mtop">facility</a> and will include a 30,000-square-foot clean room — the first of its kind designed specifically to produce power MOS semiconductors on 8-inch wafers rather than the traditional 6-inch wafers. Utilizing the larger wafer dramatically increases the yield — or number of circuits produced.	Paragraph break. Click on **facility** to go to HarrisSemi_blurb.html#mtop.
<h1><u>KEY FACTS</u></H1> <UL> <LI><FONT SIZE="5"><B>PRODUCTS : </B> MOSFet, IGBT, MCT, and Smart Discretes <LI><FONT SIZE="5"><B>CLASS : </B> 100% Auto SMIF - Class 1 <LI><FONT SIZE="5"><B>EQUIPMENT : </B> State of the Art, IC Class, Smart Tag <LI><FONT SIZE="5"><B>CAPACITY : </B> Phase I = 81,000 Wafers/year <LI><FONT SIZE="5"><B>CAPACITY : </B>Phase II = 216,000 Wafers/year <LI><FONT SIZE="5"><B>COMPLETION : </B>Phase I = January, 1997 <LI><FONT SIZE="5"><B>PROJECT LEADERS : Bill Burrell and Dave Hollock<b> </ul>	Show "KEY FACTS" in the largest heading size, and underlined. <u> begins underlining, and </u> ends it. <ul> begins an unformatted list. <li> denotes a list element. <FONT SIZE="5"> denotes a large font (standard is 3). <b> begins boldface text, and </b> ends it.
<a href="raptor_status.html"><img src="jobs.gif" WIDTH=51 HEIGHT=49 BORDER=0 alt="" >Construction Status</a><B> - UPDATED 7/19/96 	Click on the 51 by 49 pixel jobs.gif, or **Construction status**, to go to raptor_status.html. is a line break.
<a href="raptor_facts.html"><img src="data.gif" WIDTH=51 HEIGHT=49 BORDER=0 alt="">Detailed Information</a>	Click on the 51 by 49 pixel data.gif, or **Detailed Information**, to go to raptor_facts.html.

Table 13.5. *Continued.*

HTML	Explanation
`<hr><FONT SIZE="3"><b>` For more information or Mountaintop job opportunities:` ` Human Resources` ` HARRIS Semiconductor` ` 125 Crestwood Rd.` ` Mountaintop, PA 18707` ` Phone: 717-474-6761 `<p>`Other `<a href="http://www.semi.harris.com/` `employment_opportunities/">`Harris Semiconductor Employment Opportunities`</a>`	`<hr>` creates the black horizontal line. Size 3 font is normal. Click on **Harris Semiconductor Employment Opportunities** to go to http://www.semi.harris.com/employment_opportunities/.

`<hr></b>`
`<a href="welcome.html#menu">`
`<img src="home.gif" WIDTH=51 HEIGHT=49 BORDER=0 alt="Return Home"></a>`
 	
home.gif
	Click on home.gif or **Return Home** to go to welcome.html#menu.

`<a href="#top"><img src="top_btn.gif" WIDTH=51 HEIGHT=49 BORDER=0 alt="Top"></a>` top_btn.gif	Return to the page top.
` <A HREF="owner.html">`Clint Chamberlin, Webmaster`</A>`	Line break. Click on **Clint Chamberlin, Webmaster** to go to owner.html.
`<img src="copyrgt.gif" WIDTH=21 HEIGHT=19 BORDER=0 alt="Copyright">` - `<a href="mailto:cchamber@harris.com">`cchamber@harris.com`</a>` © copyrgt.gif	Show the 21 by 19 pixel copyrgt.gif. Click on **cchamber@harris.com** to send e-mail to Clint Chamberlin (mailto: command)
`<ADDRESS>` For more information about Harris Semiconductors:` ` U.S. and Canada callers dial 1-800-4-HARRIS (1-800-442-7747) ext. 700` ` International callers dial 407-727-9207` ` `</ADDRESS>`	
`<!— hhmts start —><i>`Last modified: 7/19/1996`</I><!— hhmts end —>` `</body></html>`	`<i>` begins italics, `</i>` ends italics. `</body>` ends the document body. `</html>` ends the html designation.

background and the text. For example, black text on white, or white on black, is easy to read. White or yellow on dark blue is reasonable. Blue text on black or yellow on white are poor choices. The author has seen Web pages that are very hard to read because the text color is close to the background color.

Netscape Navigator Gold 3.0 handles the background color definition automatically. The user selects a color from a menu, and Navigator writes the appropriate code.

Navigator Gold 3.0 and HTML also allow the user to define wallpaper, or a patterned background. (Microsoft's FrontPage and other Web page authoring software probably have the same capability.) The wallpaper is a .GIF file, and the browser displays it as a repeating pattern behind the page content. Again, do not select a background that makes the content hard to see. Corel PhotoPaint's color/contrast/intensity control can turn a .GIF image into a faint watermark. The watermark will show through the page content, but will not make it hard to read.

Publishing Documents and Spreadsheets

Microsoft (http://www.microsoft.com/) offers a free, downloadable Internet Assistant to go with Word for Windows (version 6a and newer). Installation of the Internet Assistant allows Word to save documents as .HTM files. There are also Internet Assistants for the Excel spreadsheet, Access database, and PowerPoint (Windows 95 version) presentation program.

Corel (http://www.corel.com/) Quattro Pro 7 can save spreadsheets as HTML files, and Corel WordPerfect 7 can publish HTML files. Corel Presentations 7 (the counterpart of PowerPoint) can generate an Internet slide show. Lotus (http://www.lotus.com/) says that "Lotus SmartSuite is the first suite to be totally Internet enabled." For example, Word Pro '96 can generate HTML files.

PowerSoft's (http://www.powersoft.com/) Formula One/NET Pro goes a step further and allows a Web browser to use a live spreadsheet. The $79 software package can convert a spreadsheet into a browser-usable format. Users with the free downloadable plug-in can then use the spreadsheet by entering their own numbers. PowerSoft's information says, "With Formula One, your applications instantly get an Excel-like spreadsheet interface with which your users are familiar. Select cells and ranges, interactively copy and move ranges of data, automatically fill cells, resize rows and columns, just like Excel. Plus, Formula One provides support for 130 worksheet functions and the Excel-style formula syntax."

One can even link Excel spreadsheets and Word documents directly. Netscape Navigator allows users to define "helper" applications for different file extensions. If, for example, the user clicks on a link to an .XLS (Excel) spreadsheet, Navigator will launch Microsoft Excel to view the document. Mountaintop has done this with its internal server. For public Web sites, the server must allow users to download files.

MathCAD 6.0 Plus can view, and use, online MathCAD (.MCD) files. An engineer or scientist could post a MathCAD document on a Web site, and it would be available to other MathCAD users. Interactive spreadsheets and programs allow Webmasters to add value to their pages and attract visitors.

Previewing Your Web Page

If you have Netscape Navigator, go to the FILE menu and select OPEN FILE. (Microsoft's Internet Explorer and other Web browsers may have different procedures.) Open your .HTM file, and Navigator will display it as it will appear on the Web. To work properly, the page must have access to any graphics (.GIF or .JPG) files that its IMG commands require. It does not need access to links (HREF=), although you will get an error message if you try to select them.

Once you've uploaded your files to the server, it is good practice to test each link and make sure that it works.

Advanced Features

Web pages can display animated images and can be interactive. So far, we've discussed Web pages that perform the following functions.

1. Create text and stationary images with .GIF or .JPG files. Both formats are compressed bitmaps.

   —.GIF is CompuServe's Graphics Interchange Format. It supports 8-bit ($2^8 = 256$) colors.

—.JPG (or JPEG) is Joint Photographic Experts Group's format. It supports 24-bit (2^{24} = 16.7 million) colors. CorelDRAW 5.0 supports JPEG images.

2. Allow the user to follow a link to another page by clicking on an image or highlighted text.

This is adequate to provide information about your business or service. For example, you can

1. Display products, with information about them.

   —Harris Semiconductor provides product data sheets online.

   —A realtor can display color photos of houses, instead of paying for magazine-type catalogs with black and white pictures. Some realtors are already doing this.

2. Display color brochures about your company.

3. Allow the user to send you e-mail by clicking on a highlighted line.

   —For example, "E-mail us at **TechSales@XYZ.com**" (all the user has to do is click on the highlighted line).

These are powerful features, but one can go further. The Java programming language can add animated images to a Web page, and VRML allows a programmer to create three-dimensional worlds. For example, an architectural firm could display building designs and let customers "walk" through them. It is also possible to interlace .GIF files to produce animated images; Alchemy Mindworks' GIF Construction Set for Windows ($20.00 + $5.00 shipping)* can create an animated .GIF from separate .GIF images. The user can specify the time between each image.

Figure 13.3 shows a training application for the semiconductor industry. The figures are drawings in Microsoft Powerpoint slides. The originals are in color, and Harris Semiconductor's Mountaintop plant uses them to teach employees how the semiconductor fabrication process works.

These can go on the Internet, or a company Intranet, as an animated slide show. The second slide (development of the exposed photoresist) would show the exposed sections disappearing in a developer solution.

---

*http://www.mindworkshop.com/alchemy/alchemy.html. The software can also make .GIF files transparent.

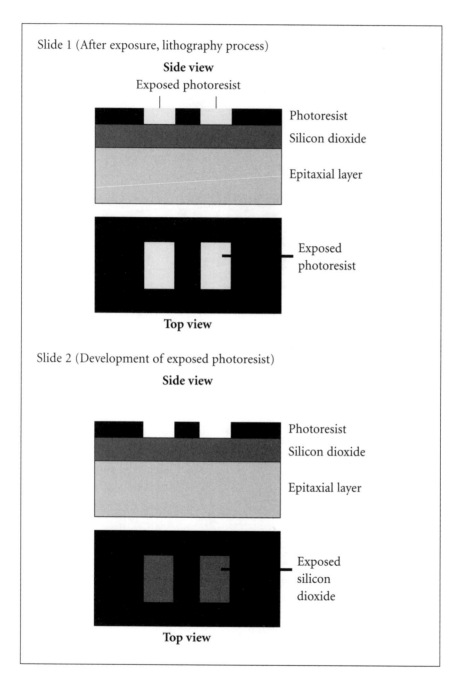

Figure 13.3. Semiconductor fabrication process (partial).

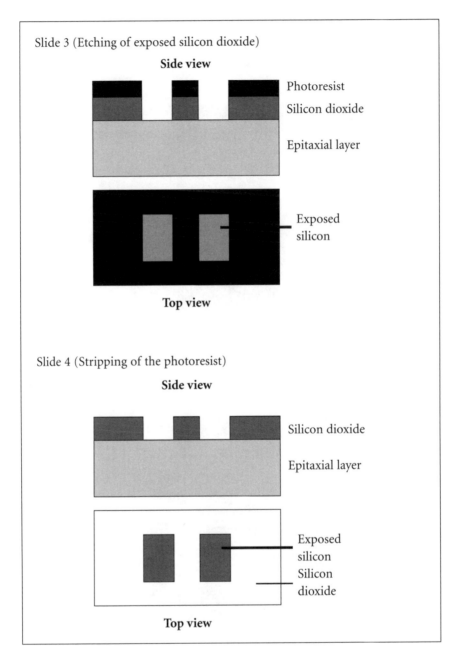

Figure 13.3. *Continued.*

The third slide (etching) would show the exposed silicon dioxide dissolving in the etchant. When hydrofluoric acid is the etchant, the openings are bowl shaped; the animation would show the bowls getting wider and deeper. If the computers have sound cards and speakers, an instructor could narrate the slides. Microsoft (http://www.microsoft.com/) offers a free program that will publish a PowerPoint slide show as an Internet slide show. It requires PowerPoint for Windows 95. Office 97 products have built-in Internet capability.

HTML supports online forms that make the page interactive. For example, an online form can accept a product order, payment information, and shipping information. Companies that accept credit card numbers must provide security against thieves who may intercept the information. Mark Twain Bank is introducing secure electronic cash, or Ecash, for Internet transactions (Templin 1996).

Adding a Web Site to Search Engines

Once you've created a Web site, you need to let people know it exists. People can't use your Web site if they can't find it. You need to let them know it's there by indexing it with the major search engines. The usual procedure is to call up the search engine (for example, Alta Vista, Excite, Lycos, WebCrawler, Yahoo) and find the option "Add URL." This adds a universal resource locator (URL) to the search engine's database. Some search engines ask for a description or keywords, while others automatically extract keywords from your main (index) page. Alta Vista is among these, and it requires META commands such as those in the earlier example. Each search engine provides instructions for adding a Web site.

Some search engines ask you to provide the URL of each page you want them to index. For example, if your Web site has 20 pages and six major ones, you should provide at least the six major links. Others, like Alta Vista, use a program called a "spider" to explore your Web site and index the links automatically.

Author Biographies

John L. Benjamin

John L. Benjamin is an engineering leader for the process development and photo and etch groups at the Harris Semiconductor facility in Mountaintop, Pennsylvania. He has been at the facility for eight years. Prior to this assignment, he was with the Semiconductor Division of the General Electric Company in various individual contributor and leadership roles. Benjamin holds a master's degree in business administration with a concentration in operations management from the University of Scranton and a bachelor's degree in materials engineering from Rensselaer Polytechnic Institute. He holds one patent and has coauthored six papers. Benjamin is the management representative for ISO 9002 at Mountaintop and served in this position during the successful certification of the Mountaintop facility in 1994.

Roger A. Bishop

Roger A. Bishop has been the manager of human resource operations at the Harris Semiconductor facility in Mountaintop, Pennsylvania, for the past 10 years. Prior to this assignment, Bishop held various human resource positions in General Electric and RCA facilities in New York, New York; Princeton, New Jersey; Boston, Massachusetts; Lancaster, Pennsylvania; and Vancouver, Washington. Bishop holds a master's degree in human resource administration from the University of Scranton (Scranton, Pennsylvania) and a bachelor's degree in advertising from

Ferris State University. He is an adjunct faculty member of the University of Scranton Graduate School of Human Resource Administration. He is also a member of the board of directors of the Greater Wilkes-Barre Area Labor Management Council and the Tri-County Personnel Association.

Michael A. Caravaggio

Michael A. Caravaggio is an engineering leader in wafer probe, calibration, and special projects. He has a degree in electrical engineering technology from Pennsylvania State University and has been involved with power discrete devices at the Mountaintop location since May 1961. He has held engineering and management assignments in equipment maintenance, calibration, equipment and process engineering, and manufacturing with RCA, GE, and Harris Semiconductor at the Mountaintop location. He provides the leadership for several diverse engineering project teams and is Mountaintop's TPM Champion, providing the TPM emphasis for the plant.

Clinton A. Chamberlin

Clinton A. Chamberlin is the manager of quality and reliability at the Mountaintop facility and has held various positions in the field of quality and reliability since joining RCA in 1973. He pioneered the use of statistical process control and real-time product reliability monitoring at the Mountaintop plant. Chamberlin holds a master's degree in engineering management from Syracuse University and a bachelor's degree in electrical engineering from Bucknell University in Pennsylvania.

Robert C. Fitch Jr.

Robert C. Fitch Jr. (M.S. electrical engineering, Air Force Institute of Technology) has been the integrated yield management project leader at Harris Mountaintop since November 1995. Prior to employment with Harris, Fitch was a reliability and quality assurance engineer with Motorola's Logic Division, Reliability and Quality Assurance Group, where his focus was on failure analysis of logic integrated circuits and developing advanced FA techniques such as emission microscopy and voltage contrast. Before joining Motorola, he was an Air Force officer for 11 years. Fitch spent three years on high power, microwave, gallium arsenide transistor development, and technology transfer to industry. He earned a bachelor of science degree in mechanical engineering from the Pennsylvania State University and a bachelor of science degree in electrical engineering from Louisiana Tech University.

Raymond T. Ford

Raymond T. Ford has been director of plant operations at the Harris Semiconductor facility in Mountaintop, Pennsylvania, for the past four years. Prior to this assignment, Ford held various manufacturing positions for RCA, GE, and Harris at Palm Beach, Florida; Syracuse, New York; and Research Triangle Park, North Carolina. Ford holds a bachelor of science degree in electrical engineering from Wilkes University, Pennsylvania. Ford has attended the Avraham R. Goldratt Institute and has earned the title "Jonah."

Jeffrey E. Lauffer

Jeffrey E. Lauffer (B.S. physics, Indiana University of Pennsylvania), has been involved with power discrete MOSFETs at the Mountaintop location since their inception during the late 1970s. He has held assignments of increasing responsibility in process engineering and management with RCA, GE, and Harris Semiconductor at this location. He was the wafer fab manager responsible for designing, building, ramping, and running the current 6″ Power MOS fab from 1987 to 1991. Lauffer is presently the engineering manager for the four manufacturing wafer fabs (4″, 5″, 6″, and 8″) and epi area at the Harris Mountaintop plant.

William A. Levinson

William A. Levinson is a staff engineer and industrial statistician at Harris Semiconductor's plant in Mountaintop, Pennsylvania. He is author of *The Way of Strategy* (1994, ASQC Quality Press) and coauthor with Frank Tumbelty of *SPC Essentials and Productivity Improvement: A Manufacturing Approach* (1997, ASQC Quality Press). A graduate of the Pennsylvania State University, he holds master's degrees in engineering, business administration, and applied statistics from Cornell University and Union College. He holds ASQ certifications in quality engineering, reliability engineering, quality management, and quality auditing. He is a registered professional engineer in Pennsylvania.

Robert F. Longenberger

Robert Longenberger has been involved in power semiconductor operations since the early 1970s with RCA Solid State Division, GE Solid State, and now Harris Semiconductor at the Mountaintop, Pennsylvania, facility. His current position entails the operational and engineering responsibility for the Calibration Laboratory in addition to manufacturing support for the 4″, 5″, 6″, and 8″ wafer fabrication areas. He is a summa cum laude graduate of Luzerne County Community College (Pennsylvania), earning an associate of science degree in electronics.

Robert E. Murphy Jr.

Robert E. Murphy Jr. has been manager of manufacturing at the Harris Semiconductor facility in Mountaintop, Pennsylvania, for the past six years. Prior to this assignment, Murphy held various manufacturing positions in Harris facilities in Palm Bay, Florida, and Poughkeepsie, New York. He holds a master's degree in organizational management from College Misericordia, Pennsylvania, and a bachelor's degree in business administration from Florida Institute of Technology. Murphy has attended the Avraham R. Goldratt Institute and has earned the title "Jonah." He is currently leading efforts within Harris Semiconductor in the use of theory of constraints and synchronous flow management applications.

Allen L. Sands

Allen L. Sands (B.S. electrical engineering, The Pennsylvania State University) has been manager of worldwide power product engineering, quality and reliability, and HiRel operations for the past five years. Prior to this assignment, Sands held various engineering and management positions with RCA, GE, and Harris Semiconductors in design, quality/reliability, product engineering, process development, and wafer fab. Sands has been the recipient of various achievement awards and holds the "Jonah" certification in Theory of Constraints from the Avraham R. Goldratt Institute.

Puneet Saxena

Puneet Saxena has been a manufacturing analyst with Harris Semiconductor for the past four years. Currently he is responsible for managing production in Mountaintop's 5″ wafer fab and epitaxial growth areas and is actively involved with the training and implementation of synchronous flow manufacturing and the theory of constraints. Puneet is a certified Jonah and a TOC Production Application Licensee from the Avraham Goldratt Institute in New Haven, Connecticut. He has been appointed as Harris Corporation's Best Practice Expert for capacity modeling in semiconductor fabrication and probe facilities. Puneet holds a B.S. degree in mechanical engineering from the Indian Institute of Technology in Kanpur, India. He also holds an M.S. degree in mechanical engineering and an MBA with a concentration in operations and logistics management from the Ohio State University in Columbus, Ohio.

Stephen E. Tetlak

Stephen E. Tetlak (currently enrolled M.S. electrical engineering, Wilkes University; B.S. electrical engineering technology, Penn State University) has been a yield enhancement engineer in the integrated yield management group at Harris Semiconductor since September 1994. His current position entails systematic failure analysis, yield analysis, and yield partitioning, of both short-loop and long-loop yield detractors used to drive defect density reduction of manufacturing facilities Fab 5, 6, and 8.

Martin L. Wentz

Martin L. Wentz (B.S. business administration, Wilkes University; A.A.S. electrical engineering technology, Luzerne County Community College) has been with Harris since 1986. Starting as a senior production supervisor, he has held positions of increasing responsibility, including an assignment at SEMATECH (Austin, Texas) as technology transfer manager. He is currently manager of training and organizational development for the Mountaintop plant. Wentz has been actively involved in team development for most of his career and has published several technical articles on the subject.

Beth Hollock, artist for Alexander and the Gordian Knot

Beth Hollock is a 13-year-old student in the Crestwood Area School District, Mountaintop, Pennsylvania. For the past five years she has received art lessons from Anita Herron. Beth received first, second, and third place awards in art for different mediums for pieces entered in school contests and the Luzerne County Fair. Beth's goal is to become an illustrator for Disney Studios.

Bibliography

AIAG. 1995. *Measurement system analysis manual; Potential failure mode and effects analysis; Production part approval process.* http://www.aiag.org.

Arter, Dennis. 1994. *Quality audits for improved performance,* 2d ed. Milwaukee: ASQC Quality Press.

Asimov, Isaac. 1958. The feeling of power. *If: Worlds of Science Fiction* (February): 4–11.

Bakker, Robert M. 1996. Why companies fail quality audits. *Manufacturing Engineering* (May).

Bales, R. E. 1950. *An interactive process analysis: A method for the study of small groups.* Reading, Mass.: Addison-Wesley.

Barker, Joel. 1993. *Paradigm pioneers.* Videotape. Charthouse International Learning Corporation, http://www. charthouse.com.

Bennis, W., and H. Shepard. 1956. A theory of group development. *Human Relations* 9:418–481.

Bishop, R. A., and R. E. Murphy Jr. 1993. The evolution of self-directed work teams within a collective bargaining environment [Summary]. *Proceedings of the 1993 IEEE/SEMI Advanced Semiconductor Manufacturing Conference, 18-25.* Library of Congress #93-78878.

Borman, Stu. 1996. Is the Web really worldwide? *Chemical and Engineering News,* 28 October, 35.

Card, Orson Scott. 1985. *Ender's game*. New York: TOR Books.

Carroll, P. 1993. *Big blues: The unmaking of IBM*. New York: Crown Publishers.

Clausewitz, Carl von (d. 1831). 1976. *On war*. Translated by M. Howard and P. Paret. Princeton, N.J.: Princeton University Press.

Cohen, Eliot, and John Gooch. 1991. *Military misfortunes: The anatomy of failure in war*. New York: Free Press.

Cooper, P. J., and N. Demos. 1991. Losses cut 76 percent with control chart system. *Quality* (April).

Courchaine, W., and K. Williams. 1992. Continuous quality improvement. Presentation at the American Production and Inventory Control Society, Mid-Hudson Chapter meeting, 11 March, Newburgh, New York.

Covey, Stephen R. 1991. *Principle-centered leadership*. New York: Simon & Schuster.

Cusimano, Jim. 1996. Understanding and using design of experiments. *Quality* (April): 78–ff.

Deal. T. E., and A. A. Kennedy. 1982. *Corporate cultures*. Reading, Mass.: Addison-Wesley.

Fechter, W. F. 1993. The competitive myth. *Quality Progress* (May): 87–89.

Feigenbaum, Armand. 1991. *Total quality control*. New York: McGraw-Hill.

Ficher, B. A., and R. K. Stutman. 1987. An assessment of group trajectories: Analyzing developmental break-points. *Communications Quarterly* 35, no. 2:105–124.

Gates, William H. 1995. *The road ahead*. St. Paul, Minn.: Penguin HighBridge Audio.

Gershenfeld, M. K., and R. W. Napier. 1993. *Groups theory and experience*. Boston: Houghton-Mifflin.

Goldratt, Eliyahu, and Jeff Cox. 1992. *The goal*. Croton-on-Hudson, N.Y.: North River Press.

Goldratt, Eliyahu. 1996. *The race*. Croton-on-Hudson, N.Y.: North River Press.

———. 1997. *Critical chain*. Croton-on-Hudson, N.Y.: North River Press.

Green, Peter. 1991. *Alexander of Macedon.* Berkeley, Calif.: University of California Press.

Hare, P., and D. Naveh. 1984. Group development at the Camp David Summit. *Small Group Behavior* 15, no. 3:299–318.

Harriott, Peter. 1964. *Process control.* New York: McGraw-Hill.

Hodge, B. J., and W. P. Anthony. 1991. *Organization theory: A strategic approach.* Needham Heights, Mass.: Allyn and Bacon.

Hradesky, John. 1988. *Productivity and quality improvement: A practical guide to implementing statistical process control.* New York: McGraw-Hill.

The inspiration for the offset press. 1995. *Harriscope,* April.

Juran, Joseph, and Frank Gryna. 1988. *Juran's quality control handbook,* 4th ed. New York: McGraw-Hill.

Kapur, K. C., and L. R. Lamberson. 1977. *Reliability in engineering design.* New York: John Wiley & Sons.

Keegan, John. 1987. *The mask of command.* New York: Penguin Books.

Kipling, Rudyard. 1940. *Rudyard Kipling, complete verse.* New York: Doubleday.

————. 1982. The rout of the white hussars. In *Rudyard Kipling, illustrated.* New York: Crown Publishers.

Larson, Carl, and Frank Lafasto. 1989. *Teamwork: What must go right, what can go wrong (Series in Interpersonal Communication, Vol. 10).* Thousand Oaks, Calif.: Sage Publications.

Levinson, W. 1994a. Multiple attribute control charts. *Quality* (December).

————. 1994b. Statistical process control in microelectronics manufacturing. *Semiconductor International* (November): 95–102.

————. 1994c. *The way of strategy.* Milwaukee: ASQC Quality Press.

Levinson, William A., and Frank Tumbelty. 1997. *SPC essentials and productivity improvement: A manufacturing approach.* Milwaukee: ASQC Quality Press.

Machiavelli, Niccolò. 1965. *The prince.* New York: Airmont Publishing.

Messina, William. 1987. *Statistical quality control for manufacturing managers.* New York: John Wiley & Sons.

Montgomery, Douglas. 1996. *Introduction to statistical quality control,* 3rd ed. New York: John Wiley & Sons.

Mountaintop team visits . . . to establish stronger lines of communication. 1994. *Harriscope,* October.

Murphy, Robert E., Puneet Saxena, and William Levinson. 1996. Use OEE; don't let OEE use you. *Semiconductor International* (September): 125–ff.

Nakajima, Seiichi. 1989. *TPM development program: Implementing total productive maintenance.* Cambridge, Mass.: Productivity Press.

O'Connell, Vanessa. 1996. Stock answer. *Wall Street Journal,* 17 June, R8.

Paterno, J., and B. Asbell, B. *Paterno: By the book.* New York: Berkley Books.

Pender, H., and K. McIlwain. 1936. *Electrical engineer's handbook: Electric communication and electronics.* New York: John Wiley & Sons.

Peters, Tom. 1987. *Thriving on chaos.* New York: Harper & Row.

———. 1988. *Structures for the year 2000.* Palo Alto, Calif.: The Tom Peters Group.

———. 1989. When surviving is not enough. Presentation to the Cornell Society of Engineers, 28 April.

Peters, Tom, and N. Austin. 1985. *A passion for excellence.* New York: Warner Books.

Peters, Tom, and R. Waterman. *In search of excellence.* New York: Harper & Row.

Reaching out to customer pays off. 1994. *Harriscope,* May.

Regan, G. 1987. *Great military disasters.* New York: M. Evans and Company.

Richards, Bill. 1996. Inside story. *Wall Street Journal,* 17 June, R23.

Rogers and Ferktish. 1996. Creating a high-involvement culture through a value-driven change process. Development Dimensions International, www.ddiworld.com.

Sandberg, Jared. 1996. Making the sale. *Wall Street Journal*, 17 June, R6.

Savola, Tom. 1995. *Using HTML*. Indianapolis, Ind.: Que Corporation.

Schermerhorn, J. R., J. G. Hunt, and R. N. Osborn. 1985. *Managing organizational behavior*, 2d ed. New York: John Wiley & Sons.

Scholtes, Peter R. 1988. *The team handbook*. Madison, Wis.: Joiner Associates.

Schultz, W. C. 1966. *FIRO: A three-dimensional theory of interpersonal behavior*. New York: Holt, Rinehart and Winston.

Scott, W., and D. Hart. 1990. *Organizational values in America*. New Brunswick, N.J.: Transaction Publishers.

Scotto, Michael J. 1996. Seven ways to make money from ISO 9000. *Quality Progress* (June): 39–41.

Shirose, Kunio. 1992. *TPM for operators*. Cambridge, Mass.: Productivity Press.

Smith, Ralph. 1976. *Circuits, devices, and systems*. New York: John Wiley & Sons.

Spence, Gerry. 1995. *How to argue and win every time*. Audiotape. Los Angeles: Audio Renaissance Tapes.

Struebing, Laura. 1996. Customer loyalty: Playing for keeps. *Quality Progress* (February): 25.

Sun Tzu. 1963. *The art of war*. Translated by Samuel Griffith. New York: Oxford University Press. [Original version: ~500 B.C.E.]

———. 1983. *The art of war*. Translated by James Clavell. New York: Delacorte Press.

Templin, Neal. 1996. Cash crunch. *Wall Street Journal*, 17 June, R22.

Tuckman, B. W., and M. A. C. Jensen. 1977. *Stages of small-group development revised*. *Group and Organizational Studies* 2, no. 4:419–427.

Twain, Mark. 1989. *A Connecticut Yankee in King Arthur's court*. New York: Harper & Row.

Wiener, Leonard. 1996. How I bought my new car—Online. *U.S. News and World Report*, 6 May, 75.

Zachary, G. Pascal. 1996. Hard labor. *Wall Street Journal*, 17 June, R26.

Index